CAMBRIDGE LIBRARY COLLECTION

Books of enduring scholarly value

Life Sciences

Until the nineteenth century, the various subjects now known as the life sciences were regarded either as arcane studies which had little impact on ordinary daily life, or as a genteel hobby for the leisured classes. The increasing academic rigour and systematisation brought to the study of botany, zoology and other disciplines, and their adoption in university curricula, are reflected in the books reissued in this series.

The Geology of the Island of Arran

The Scottish geologist Andrew Crombie Ramsay (1814–91), best remembered for his work on glaciation, made his name with this study, which originated in holiday visits to Arran. Encouraged by John Nichol of Glasgow University, Ramsay had prepared a geological map of the island for the British Association's visit in 1839, and was to have led a field excursion and lectured to the delegates there, but missed the boat. Nichol arranged for Ramsay's work to be published the following year. It attracted the attention of Roderick Murchison, who found him employment with the Geological Survey, and Ramsay later succeeded Murchison as its director and was knighted on his retirement in 1881. The book, designed as a practical visitor's guide for both amateur and professional geologists, is organised by district with reference to specific geological features, and Ramsay provides full explanations, diagrams and engravings to make his work accessible to non-specialists.

Cambridge University Press has long been a pioneer in the reissuing of out-of-print titles from its own backlist, producing digital reprints of books that are still sought after by scholars and students but could not be reprinted economically using traditional technology. The Cambridge Library Collection extends this activity to a wider range of books which are still of importance to researchers and professionals, either for the source material they contain, or as landmarks in the history of their academic discipline.

Drawing from the world-renowned collections in the Cambridge University Library, and guided by the advice of experts in each subject area, Cambridge University Press is using state-of-the-art scanning machines in its own Printing House to capture the content of each book selected for inclusion. The files are processed to give a consistently clear, crisp image, and the books finished to the high quality standard for which the Press is recognised around the world. The latest print-on-demand technology ensures that the books will remain available indefinitely, and that orders for single or multiple copies can quickly be supplied.

The Cambridge Library Collection will bring back to life books of enduring scholarly value (including out-of-copyright works originally issued by other publishers) across a wide range of disciplines in the humanities and social sciences and in science and technology.

The Geology of the Island of Arran

Andrew Crombie Ramsay

CAMBRIDGE UNIVERSITY PRESS

Cambridge, New York, Melbourne, Madrid, Cape Town,
Singapore, São Paolo, Delhi, Tokyo, Mexico City

Published in the United States of America by Cambridge University Press, New York

www.cambridge.org
Information on this title: www.cambridge.org/9781108037778

This edition first published 1841
This digitally printed version 2011

ISBN 978-1-108-03777-8 Paperback

The original edition of this book contains a number of colour plates, which have been reproduced in black and white. Colour versions of these images can be found online at www.cambridge.org/9781108037778

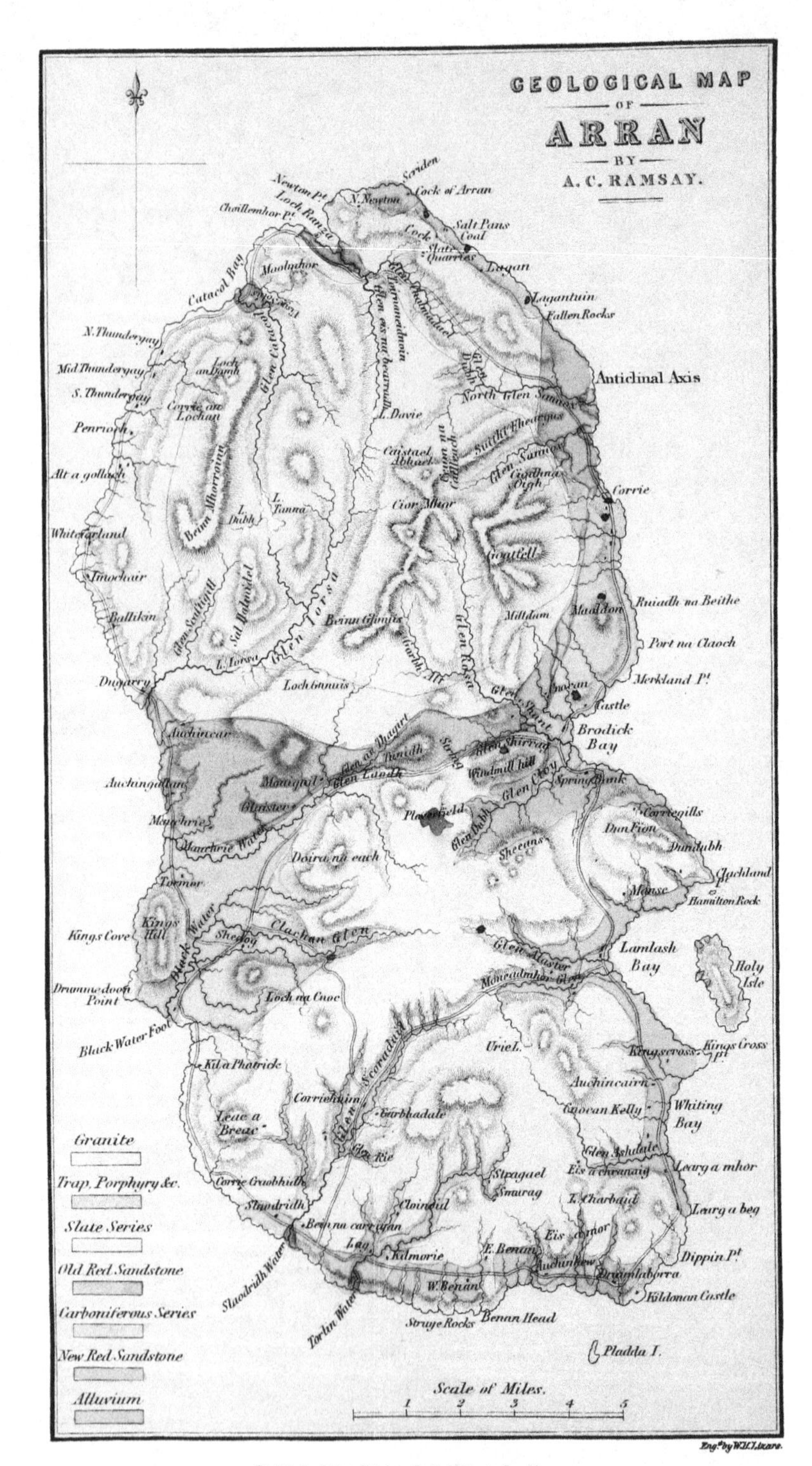

Published by Richard Griffin & Co. Glasgow.

THE GEOLOGY OF THE ISLAND OF ARRAN, FROM ORIGINAL SURVEY.

BY

ANDREW CROMBIE RAMSAY.

ILLUSTRATED BY ENGRAVINGS.

GLASGOW:
PUBLISHED BY RICHARD GRIFFIN & COMPANY,
AND THOMAS TEGG, LONDON.

MDCCCXLI.

TO

J. P. NICHOL, LL.D., F.R.S.E.,

PROFESSOR OF PRACTICAL ASTRONOMY IN THE UNIVERSITY OF GLASGOW,

THIS WORK

IS RESPECTFULLY INSCRIBED.

PREFACE.

THE following sketch has been executed with the view to afford a useful GUIDE to "The Geology of Arran;" and as many geologists who visit that Island, do so with the intention of becoming practically acquainted, for the first time, with the more striking phenomena of the subject, it has been deemed expedient frequently to explain and illustrate many well known geological phenomena which the experienced geologist might consider to be of themselves sufficiently obvious. At the same time, the work may not perhaps be found useless as a guide by the practical geologists who so frequently visit that interesting Island, for the purpose of making personal observations on phenomena, which they have not so favourable an opportunity of examining in other districts.

In writing a treatise on the geology of Arran, the most obvious method would be to treat of all its different formations according to their relative ages, without, at the same time, referring to the various formations of any particular district; and any one previously acquainted with Arran would most easily understand a description of its geology thus arranged. But as this work is principally intended as a guide over the Island, it was necessary to divide the country into districts, and occasionally, and more particularly in the concluding chapter, to generalise and sum up the facts, which are brought, in a detached manner, under the observer's notice, as he proceeds over the face of the country.

The author has to acknowledge his obligations to RODERICK IMPEY MURCHISON, Esq., and the Rev. ADAM SEDGWICK, for the use of their admirable Section (No. I.) of the Coast of Arran, from Loch Ranza to Clachland Point,—a section which cannot be too much praised for its perfect accuracy and distinctness. He has also availed himself of PROFESSOR JAMESON'S drawing and description of the remarkable igneous veins that exist in the neighbourhood of Tormor.

The geological survey, as delineated on the map,* is entirely the product of original and personal observation; and though it may probably exhibit some minor errors, it is confidently hoped that the disposition of the various formations, on the whole, will be found correct. The topographical details are copied from the large map of Arran, executed from the survey of MR. BAUCHOP.

The author also begs to express his acknowledgments to JOHN PATERSON, Esq., Lamlash, and to the other gentlemen and inhabitants of Arran generally, for the facilities which they afforded him while conducting his surveys.

GLASGOW, *March* 31*st*, 1841.

* A geological Model of the Island of Arran, on a scale of two inches to the mile, is now published, and may be advantageously studied in conjunction with this work.

CONTENTS.

EXPLANATION OF THE SECTIONS.

SECTION I.

SECTIONAL VIEW OF THE COAST OF THE ISLAND OF ARRAN, FROM CLACHLAND POINT TO LOCH RANZA.

According to the Survey of R. I. Murchison, Esq., and the Rev. A. Sedgwick.

From Loch Ranza to Alt Beithe, Chlorite Schist,—dip. 70° S.E.

At Alt Beithe there is a small patch of New Red Sandstone, resting unconformably on the Schist.

The New Red Sandstone and Conglomerate, stretches from Alt Mhor to the south of the Cock of Arran.

From the termination of the New Red Sandstone, the Coal Measures extend to near Croggen Point. Its beds succeed each other in regular order, without any breaks or faults, its layers of Limestone are therefore of different ages. The various alternations are particularly noticed in Chapter Third.

Immediately to the north of Croggen Point are six beds of Trap, &c.—See Chapter Third.

The Old Red Sandstone, extends from about Croggen Point to the March of Achab Farm, near Corrie.—See Chapters Second and Third.

Anticlinal Axis.—See Chapter Third.

Behind the Old Red Sandstone, the Carboniferous Series, and the New Red Sandstone, are the Unconformable Strata of Schist and Slate.

From the Old Red Sandstone to Maoldon, the interior low hills are principally Alluvial.

Between the Old Red Sandstone and the south side of Glen Shirrag, Brodick Bay, the Limestone of the Carboniferous Series, is elevated to the surface four times. They are all originally the same bed forced up in different localities, and divided by faults, indicated in the Section, (see Chapters First and Second;) thus differing from the Limestones of the Northern Carboniferous Series, each of which is a distinct and separate bed.

From Brodick Bay to Clachland Point, are Strata of New Red Sandstone, penetrated by Dykes of Trap, Pitchstone, &c. Behind and above the Sandstone Strata, are masses of Porphyry, Syenite, &c.

SECTION II.

INTERIOR SECTION OF THE ISLAND OF ARRAN, FROM THE COCK OF ARRAN TO BENAN HEAD.

According to an Original Survey by A. C. Ramsay.

1. Granite of the Interior.
2. Slate and Schistose Rocks resting on Interior Granite.
3. Old Red Sandstone resting on Slate, &c.
4. Carboniferous Rocks of Brodick Bay resting on Old Red Sandstone, and of the Interior and Southern District underlying the Southern Igneous Formations.
5. New Red Sandstone resting on the Slaty rocks of the Northern District, and on the Carboniferous Rocks of Brodick Bay, and underlying the Southern Igneous Formations.
6. Various Traps, Porphyry, Syenites, &c., which have burst through, and overlie the Southern Carboniferous Rocks, and the New Red Sandstone.
7. Fine Granite of Ploverfield associated with the Southern Igneous Rocks.
8. Trap and Pitchstone Dykes, &c., penetrating the Coarse Grained Granite.
9. Quartz Rock—Altered New Red Sandstone.

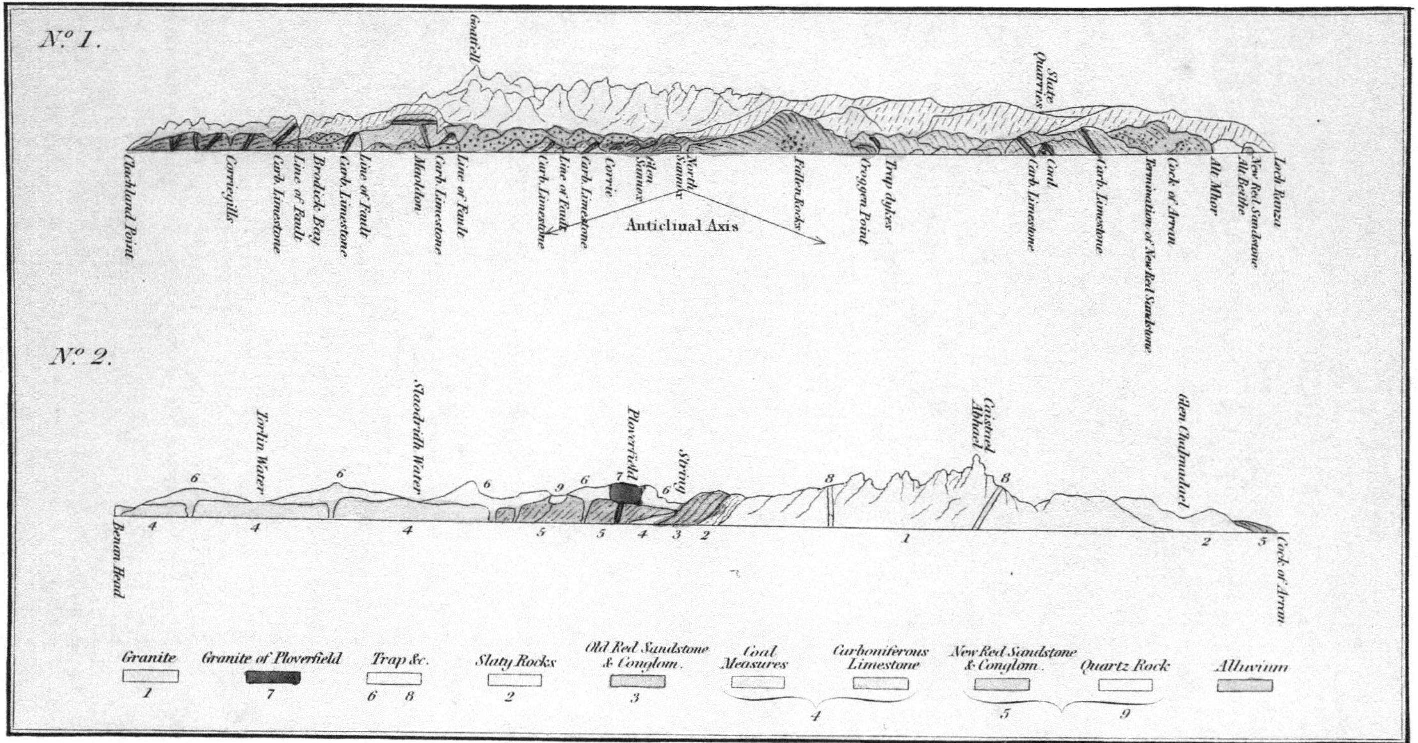

No. 1.

Loch Ranza
New Red Sandstone
Alt Beithe
Alt Mhor
Cock of Arran
Termination of New Red Sandstone
Carb. Limestone
Slate Quarries
Coal
Carb. Limestone
Trap dykes
Croggen Point
Fallen Rocks
Anticlinal Axis
North Sannox
Glen Sannox
Corrie
Carb. Limestone
Line of Fault
Carb. Limestone
Goatfell
Line of Fault
Carb. Limestone
Maoldon
Line of Fault
Carb. Limestone
Brodick Bay
Line of Fault
Carb. Limestone
Corriegills
Clachland Point

No. 2.

Cock of Arran
Glen Chalmadael
Caistael Abhael
String
Ploverfield
Slaodridh Water
Torlin Water
Benan Head

Granite 1
Granite of Ploverfield 7
Trap &c. 6 8
Slaty Rocks 2
Old Red Sandstone & Conglom. 3
Coal Measures
Carboniferous Limestone
4
New Red Sandstone & Conglom. 5
Quartz Rock
Alluvium
9

Engd. by W. H. Lizars

Published by Richard Griffin & Co. Glasgow.

View from the top of Goatfell.

CHAPTER I.

APPROACH TO ARRAN—BRODICK BAY—THE ASCENT OF GOATFELL—STRUCTURE OF THE GRANITES—CYCLOPEAN WALLS—PART OF GLEN ROSA—THE FORMATION OF VALLEYS—THE CARBONIFEROUS FORMATION OF BRODICK BAY—THE DENUDATION OF GLEN SHIRRAG—OVERLYING IGNEOUS FORMATIONS OF BRODICK BAY, AND GRANITE OF PLOVERFIELD—QUARTZ ROCK—PITCHSTONE VEINS—NEW RED SANDSTONE—CORRIEGILLS—ALLUVIAL PLAIN OF BRODICK.

The greater part of the Island of Arran, as seen from the coast of Ayrshire, presents the appearance of a picturesque mass of rugged mountains, rising bleak and bare directly from the sea, and seemingly unadorned by vegetation, and totally incapable of cultivation; but, as the spectator nears the shore, its aspect gradually softens, and milder scenes are seen to intermingle with that barren grandeur, which in truth still forms the main characteristic of the landscape. Cultivated inclosures adorn the sloping banks which approach the sea; and here and there, spread over the uplands, in the glens, and on the winding shores, wreathes of blue smoke curling on the hill sides, give token of the scattered cottages of the peasantry.

To him who has examined the structure of Arran, this distant prospect is exceedingly interesting, embracing as it does at one view, all the great geological features of the island. He will then easily observe the extent of the principal divisions of the rocks, and almost fancy that he can determine the bounds of the different formations, by the shape of the hills, and the colour of their various vegetations. To the uninitiated geologist, who has read of Arran as an epitome of the geology of the globe, the prospect is full of expectation; and small as the island is, he approaches it lost in wonder at the vastness of the operations, that heaved into being those now venerable hills, which intersect the intricate hollows on their sides, and bound, as with a wall, the dim glens which penetrate into the heart of the mountains. As the visitor enters Brodick Bay, the scene becomes exceedingly beautiful. The lofty precipices, and gloomy shadows of the rugged ridge of Ben Ghnuis, which often throws a twilight hue over the deep hollow of Glen Rosa, and strongly contrasts with the open and swelling character of the hills around Glen Cloy; the cliffs of Corriegills, the white and sloping beach which rounds the bay, the embattled castle towering above its surrounding woods, the green enclosures, and beyond these, the long expanse of brown heath, from which rises the grey peak of Goatfell,—all these form a scene of surpassing beauty, such as cannot be excelled by the most romantic scenery of the far-famed Firth of Clyde.

The great variety of geological phenomena, contained within the limited compass of this interesting island, peculiarly adapt it for the early study of those who desire to know something more of geology than what may be gained in cabinets: and the disposition of the various stratified formations, their relations to each other, and to the subjacent and superincumbent igneous rocks, altogether form an excellent school, for the commencement of the labours of the practical geologist. The beginner, who knows something of the true principles of the science, but who has not yet much observed for himself, will there find, that a few weeks' actual experience, will be of infinite service to him, by enabling him to put those principles in practice, which he may understand theoretically, but which he cannot well apply without personal observation; and he who is conversant both with the theory and practice of geology, will find much within its bounds fitted both to interest and instruct him.

This work being principally intended as a Guide to "The Geology of Arran," we shall make Brodick Bay, the principal resort of visitors, the starting place; and thence endeavour to conduct the observer on a geological tour, embracing all the more interesting features of the island. And as in so doing, it will be necessary to describe the various appearances, step by step as they occur, it will be better, not always to enter at the time on the proofs, by which the various strata are referred to the different formations; as it is impossible always to do this, without embracing the co-relative proofs, which can only be drawn from a survey of extensive tracts, that, without pre-examination, ought not at first to be described. The reader must, therefore, in the first instance, frequently take it for granted, that the classification is correct; and he who actually uses it as a guide, must be content to wade through all the dry, and apparently trifling details, which a minute description renders necessary. In the last chapter will be found those proofs and generalisations, previously omitted, which will enable him to connect the observations of facts, scattered throughout the preceding pages.

The excursion which naturally first engages the attention of the geologist, is the ascent of Goatfell. The elevation of this mountain has been variously estimated. It is generally stated to be 2,865 feet; while in the summer of 1839, as barometrically measured by M. Necker, it is stated to be 902 French Metres, equal to 2,959 English feet. The ordinary path, is by a narrow track to the eastward of Cnocan burn; but the best geological rout is in the course of that stream, in which alone, the appearances of the strata which recline against the mountain can be accurately observed. Ascending from the plain of Brodick, the water has cut a channel through the alluvial banks, which, now covered with wood, overhang the bed of the stream. It is frequently choked with boulders of granite, which have rolled from the neighbouring heights, and its bed is here composed of old red sandstone, dipping in a southerly direction, of a fine grain and dark colour, and exhibiting occasional thin transverse veins of carbonate of lime, possibly the result of infiltration into ancient cracks. At the distance of a few hundred yards above the wood, the strata assume a more finely laminated and slaty structure. They are sometimes slightly contorted, and the inclination of the strata is partially reversed, dipping N.W. towards the mountain at an angle of about 60°. The cause of this irregularity is soon discovered, for about two hundred yards further up, and immediately below the first natural clump of birch trees, a broad greenstone dyke bursts through between the strata of old red sandstone, (here a conglomerate,) and the underlying clay slate, and crossing the bed of the stream from N.W. to S.E., is lost in the overhanging alluvium. In the middle of this dyke, and in the very bed of the stream, there is a small mass of porphyritic trap, containing beautiful crystals of glassy felspar. As its base is lost in the pool, it is difficult to determine, whether it be a detached and transported block, or a later dyke penetrating the common greenstone.

Proceeding further up the stream, the strata of clay slate are sometimes almost vertical; and as it approaches the granite, owing to the action of heat, which, it is presumed, formerly proceeded from the granite, and other agencies afterwards to be explained, the stratification becomes much contorted. The layers are not, as heretofore, superimposed on each other in straight parallel lines, but are twisted into a variety of the most tortuous windings, such as the layers could not possibly have assumed during the process of stratification, and which must therefore, be referred to some other independent agency. These appearances increase, as the strata approach the base of the cone of Goatfell; till at the mill dam, the junction of the granite and slate takes place. The absolute junction of the two masses of rocks is not here visible, but that it is in the immediate neighbourhood, probably in the bed of the dam, is clearly shown by the appearance of a granite vein, about one foot broad, which penetrates the strata, and crosses the bed of the stream about ten yards below the artificial wall which confines the water of the dam; thus indicating its intrusion, while in a state of fusion, into the stratified deposit with which it came in contact. The granite is of a yellowish colour, and fine grained and compact in its texture. As a general rule, it will almost invariably be found, that wherever the granite joins the slate, and more particularly when it penetrates it in veins, it assumes a fine compact texture, and consists principally of felspar. It thus presents a totally different appearance to the coarse crystalline granite, which frequently forms the mass of the rock within a few yards of the junction.* The slate is exceedingly tortuous where it is penetrated by this vein; and the strata are intermingled with numerous veins of quartz, of varying sizes, and which generally alternate with the slaty strata in regular minute laminæ. The phenomena here described will perhaps be better understood by a reference to figure 4.

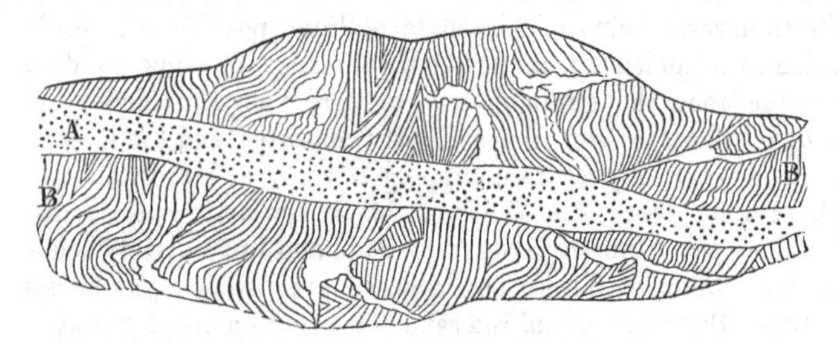

Fig. 4. Granite vein penetrating contorted strata of slate at Brodick Mill dam, Goatfell. A Granite. B Slate.

From the foregoing remarks it will be evident to the observer, that though at a higher elevation than the old red sandstone, yet the slate occupies an inferior geological position; and the granite, though topographically the highest of all the formations, is geologically the lowest of the series. For, as the slate and the old red sandstone dip in the same direction, and as the strata of the former rise from under the old red sandstone, it is at once evident that the deposition of the sandstone took place subsequent to the formation of the slate, both of them being upheaved into their present elevated and angular position, by the protrusion of the mass of granite which underlies both. This will be at once understood by reference to section No. 2, in which the relative

* It will afterwards be seen that a large tract of fine grained granite also exists in the interior. Its peculiar organization cannot, however, be affected by the causes now noticed.

positions of similar strata are thus laid down. The proof derived from superposition, which has just been applied to demonstrate the superior antiquity of the slate to the sandstone formation, does not, in this instance, apply to the granite, which, being a rock of igneous origin, and formed *under* the stratified formations, might consequently be either older or newer than the superimposed stratified formations, according as it had been formed and consolidated, previous or subsequent to the deposition of the rocks which recline on it. In the former case it would be older than the stratified rocks, but when it has been formed by the melting of still older materials, which thus have taken a new form, it is said to be of later formation than the superincumbent stratified formations through which it is protruded. The granite of Arran is of later origin than any of the stratified formations which surround it. The proofs of this will be fully entered into in the concluding chapter, when the results of the previous observations are summed up and generalised.

From the mill dam the remainder of the ascent of Goatfell is first over a deep peat moss; and where the ascent becomes more steep, over the native rock and the barren soil, which, derived from the disintegration of the granite, yields but a meager vegetation, and affords a scanty subsistence to the few sheep which feed on the mountain side.

The mass of the granite of Goatfell is the large grained variety; but there are found in it many veins of a fine compact texture, which indeed are common throughout all the coarse grained granite of Arran. These veins, it will afterwards be shown, are of later origin than the coarse granite, and probably the result of fresh irruptions which were intruded into fissures in the older rock.

There are many varieties of the coarse grained granite, but as they all belong apparently to the same period, and probably owe their origin to the same cause, it is unnecessary to particularise every minor variety. The following remarks are generally applicable to this species, wherever it is found in the island, and can be referred to in any locality that may be examined. Its constituent parts are felspar, quartz, and mica; and the differences in texture are owing to the variable proportions and sizes of these constituent minerals. Generally the felspar predominates, next quartz, and lastly mica, in comparatively small quantities. There are three varieties of felspar—the first of a light brown colour, the second almost pure white, and the third, commonly called glassy felspar, exhibiting a brilliant polish when fractured. These, though sometimes of the same colour as the quartz, can always be distinguished from it, as the felspar can be scratched with the point of a knife; whereas the quartz, from its extreme hardness, receives no impression from the knife. The first and second varieties sometimes occur in small amorphous masses, and sometimes, though rarely, in clusters of crystals.

Of the quartz, there are also several varieties: viz. common white quartz, pale yellow, light grey, colourless quartz, or pure rock-crystal;

light brown, dark brown or smoked quartz, and sometimes, though rarely, it is almost black. The common white quartz sometimes occurs in considerable masses, an example of which may be seen in a granitic boulder a little to the east of Brodick mill dam.

The transparent and dark varieties are by no means rare, and frequently attain to a considerable size. They are usually found in the form of hexagonal prisms, and occur along with crystals of felspar, in those rounded cavities which are so common in this granite. Specimens are also found, in which the transparent and light grey varieties alternate in layers, like the various colourings of an agate. The mica is much more sparingly diffused than the other minerals, and is usually found in small scales of a dark brown or black colour.

Though the fine grained granite does not occur in mass in the neighbourhood of Goatfell, being only found in the form of penetrating veins, yet in treating of the subject it may here be remarked, that it also exhibits considerable variety of structure, on account of the varying compactness of its texture, the fineness of its grains, and the relative proportion of its constituents. So much is this the case, that hand specimens might, in some instances, easily be mistaken for a harsh sandstone. Large regular crystals are seldom if ever found in it. Wherever it penetrates the coarse granite in veins, there it exhibits the finest and most compact texture, and frequently the quartz and mica totally disappear, leaving the remaining constituent in the form of a compact felspar. The phenomenon of its superior fineness in the veins is easily accounted for, when it is considered, that when the melted matter forced itself into the fissures in the coarse grained or older granite, and thus came in contact with a cooling substance, (the coarse granite being previously consolidated,) it would crystallize more rapidly, and consequently in smaller crystals, than if the cooling process were more gradual.

In this granite, mica is only to be found in very small black scales, sparingly intermingled with the quartz and felspar. The two latter constituents are sometimes mingled in nearly equal proportions, but very frequently the felspar predominates.

Near the summit of Goatfell, and also on the south shoulder, the granite suddenly rises in perpendicular cliffs, assuming the artificial appearance of huge Cyclopean walls, as exhibited in figure 5. Large

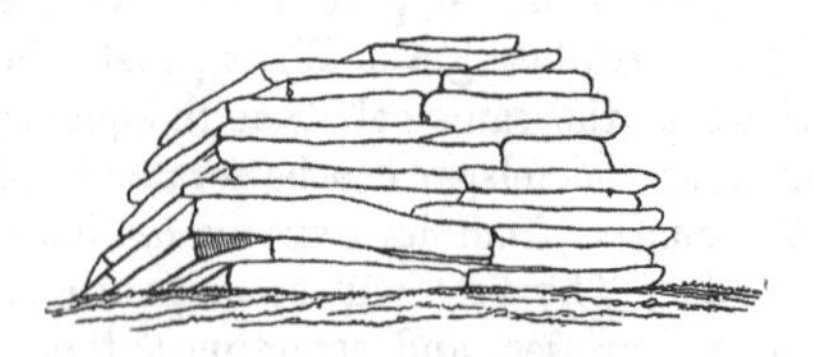

Fig. 5. Cyclopean Granite Walls.

blocks are arranged one above another with the utmost nicety, thus frequently presenting a vertical face of rock of considerable height. Such appearances are very common on the shoulders and at the summits of granite mountains. Besides those on Goatfell, they are particularly marked on the west side of Beinn Ghnuis, and on the peak of Caistael Abhael,* where several isolated masses of this description rise from the summit of the hill, and exhibit on all sides vertical cliffs, probably one hundred feet high. This, as already stated, may be seen in a less degree on Goatfell, and it is evident, that they now remain elevated above the surrounding rocks, owing to their superior hardness having offered a greater resistance to the action of the elements, which has worn away the softer surrounding masses.

The sides also of part of the south shoulder of Goatfell, along with many other districts, frequently exhibit a somewhat similar phenomenon, the granite being split into masses of varying sizes, superimposed on each other, and dipping in the direction and at the angle of the descent of the hill, like layers of lava which have flowed over each other, the one being cooled before the eruption of the superincumbent bed. But in the case of the granite, this could not be the case, as it will be afterwards shown that it was formed and consolidated previous to its protrusion through the overlying stratified formations, and consequently no such overflowings could have occurred. It is difficult to assign a reason for this phenomenon, but it seems possible that the melted mass, coming in contact with a cooling substance, viz., the slate, at its surface, refrigerated by degrees downward to the centre, and, as it thus consolidated and contracted, the different layers may have split off from the more heated mass below, in the form which they now assume.

Having reached the highest point of Goatfell, the eye of the geologist suddenly rests on a scene, which, if he be a true lover of nature, cannot fail to inspire him with astonishment and delight. The jagged and spiry peaks of the surrounding mountains,—the dark hollows and deep shady corries into which the rays of the sun scarce ever penetrate,—the open swelling hills beyond,—the winding shores of Loch Fine, and the broad Firth of Clyde studded with its peaceful and fertile islands,—the rugged mountains of Argyllshire, and the gentle curves of the hills of the Western Isles, their outlines softened in the distance, form a scene of most surpassing grandeur and loveliness. In all its varying aspects, it is a scene the memory of which can be dwelt on with pleasure; whether it be seen in the early morning, when the white mists, drawn upwards from the glens, float along the hills, and half conceal their giant peaks;—or in the gloom of an autumn evening, when the descending clouds, urged onwards by the blast, flit swiftly across the mountain sides, while ever and anon their gloomy shoulders loom largely through the rolling masses, and seem to the beholder to double their vast proportions;—or in the mellow light of a summer sunset, when the shadows of the hills fall far athwart the landscape, and the

* Caistael Abhael, or the Fortress of the Ptarmigan, so called from its tower-like summit being frequented by these birds.

distant Atlantic gleams brightly in the slanting rays of the setting sun; while, as he sinks below the horizon, it is difficult to distinguish the lofty summits of Jura and the Isles, from the gorgeous masses of clouds among which he disappears.

Having descended Goatfell, and again reached the mill dam, it is exceedingly interesting to follow the boundaries of the granite and slate to Glen Rosa, where the effects produced by the junction of the igneous and stratified rocks are remarkably well developed. Around this part of the base of the cone of Goatfell, the precise line of junction is concealed by the deep moss and heather; but as the north side of the shallow hollow which stretches from the mill dam to below Rosa, is of granite, and the mounds on the south of slate, it is evident that the boundary must be somewhere in this hollow, and for reasons which will afterwards be explained, this boundary is probably immediately below the southern rise. But where the hill begins to descend into Glen Rosa, the granite and slate may be seen within a few feet of each other, the slate being penetrated by a small vein of granite, about six inches broad. Still farther down, the example already alluded to is laid bare on the surface of a small rocky mound. Here the clay slate exhibits all the marks of having been convulsed by some powerful agency; the strata, which at first were formed in parallel layers, being curved and contorted in the most fantastic forms. The intrusion of a large granite vein into the slate, which suddenly shoots out into several smaller root-like veins, seems to indicate that the contact of the melted granite caused these singular contortions in the slate, which, cracking with the heat, admitted those veins now intersecting the strata, in the manner indicated in figure 7, page 15.

It may appear unnecessary to enlarge further on this subject, as there are few, if any, who still adhere to the old opinion of the aqueous origin of granite; but the student to whom these phenomena are new, cannot too much familiarize his eye with them; and he ought to remember that long after Hutton, who first observed them, had made the great discovery of the penetrating veins in Glen Tilt, many geologists still pertinaciously adhered to the now exploded doctrine above referred to, and thus originated that interminable war of words between the Neptunists and Vulcanists, which, like many other controversies, served rather to retard than advance, the science which the disputants professed to advocate.

It would be a needless repetition minutely to trace the line of junction into the hollow of the glen. It is however denoted by a line of slaty mounds, which mark the descent of the hill in a N. E. and S. W. direction. Above the turn of the glen, it is bounded on both sides by granite, and immediately below, where the torrent of Garbh Alt (or the rough water) joins the Rosa, the granite disappears, the strata being composed of slate dipping in a southerly direction at an angle of about 60°. The strata being thus upturned on their edges, the entire mass is seen to be of great thickness. It is difficult to conceive the period of time requisite for the deposition of a mass which, probably, on the S. W. side of the glen, may be at least 1500 feet thick.

Descending the glen, the slate is succeeded by the old red sandstone, which has already, in describing the ascent of Goatfell, been noticed as occupying part of the bed of the Cnocan Burn, from whence it stretches entirely across the Island, in the form of the common red sandstone, or in that of a conglomerate, the aggregated pebbles being sometimes almost a foot in diameter. It here forms part of the hills on both sides of Glen Rosa.

A curious phenomenon occurs in the wood immediately behind Glen Rosa Cottage, a little off the west side of the road. This rock was formerly partially quarried for the purpose of building enclosures. Part of the stone being removed, the exposed surface presents a reticulated appearance, like the fibrous stalks of a gigantic leaf, the slightly elevated reticulations, and the intervening interstices, varying in the fineness of the grain.

This appearance possibly originated in the elevation of the sand which forms the rock, previous to its consolidation. It may then have cracked in the dry atmosphere, these cracks being subsequently filled with a foreign substance now constituting the reticulations, after which the entire mass was probably again submerged and consolidated.* The same phenomenon may often be seen in any muddy pool, which is liable to occasional drainage.

The business of the geologist extends beyond a mere observation of the phenomena of rocks, and during this excursion, sufficient for a long summer day's work, he may profitably have directed his attention to the effects produced on the form of the ground by the action of running water, as evinced in Glen Rosa and the Cnocan Burn.

Every one must have remarked the powerful agency of water in the formation of ravines and valleys. The Cnocan Burn affords a good example of this on a small scale. Before the formation of the ravine through which the water now flows, an undulating hollow probably existed in the same locality. A drainage of the surrounding country thus ensued, and the accumulated waters gradually formed for themselves a channel, by which they flowed into the sea. The process would then proceed as follows:—The erosion of the water gradually wears away the alluvium or rock over which it flows, thus increasing the depth of the channel; and as the stream meets with some opposing object which stems its current, it is gradually turned from part of its original course, and wearing away the opposite bank, widens the ravine through which it flows. But one turn in a rapid stream almost invariably causes the formation of another; for as the current of the stream is reflected off the bank against which it has been forced, it again rushes

* The best example of it may be seen in a large slab of stone, which forms one of the steps at the west end of the rude bridge that crosses the Rosa below the farm-house. This stone was pointed out to me by Professor Nichol, which first induced me to search for the rock in *situ*.

against the opposite bank, which, crumbling before its force, is gradually washed away by the current. In this manner the hollow slowly deepens and increases in width; of this the Cnocan Burn affords a good example; and small as the stream is, if no other agency interfere, a deep and a wide glen may yet be excavated where it now runs.

The formation of those alluvial tracts, called in Scotland "haughs," arises from the same cause; the haugh increasing in size at the extreme point of the bend, where the water is always shallower, and the current slow, owing to the gradual destruction of the concave curve of the opposite bank, where the stream is always deepest and most powerful.* These appearances are frequently well developed in the course of the Rosa. It is evident, also, that the hollow of the glen is gradually increasing in depth; and from the appearance of the surrounding banks, it is very probable that the alluvial mound on which now stands the farm house of Glen Shant, was at some distant period on a level with the bottom of the glen.

At various places, considerably above Glen Shant, isolated alluvial mounds stand prominently out in the middle and at the sides of the glen, and are evidently the remains of ancient banks, which, from the circumstances attending the changes of the direction of the current, have been spared by the stream, while the other banks of the same antiquity, or those of later date, have frequently been destroyed. They have, therefore, thus attained a considerable height, and now denote the ancient level of the bed of the stream.

In the woods in the vicinity of the castle are two beds of limestone, (Section No. 1,) which, with their accompanying beds of sandstone and shale, belong to the carboniferous formations. The limestone also appears in the small stream which passes Brodick Inn,—at the church, and on the north side of the Windmill Hill, (Section No. 1,) about half-a-mile up the road. As this limestone, with its accompanying fossils and alternating beds of shale, are of the same nature, and belong to the same geological formation as the quarry at Corrie, where it is extensively wrought, and where it can be much more effectually examined, it is unnecessary here to enter on further particulars regarding it. It is a member of the series of carboniferous rocks which extends between Corrie and Brodick, and rests conformably on the old red sandstone. The junction of these two formations, is concealed around the base of Goatfell, and in the plain of Brodick, by accumulations of moss and alluvial deposits; but as in other instances, where visible, the transition is probably gradual.

The lower beds of the carboniferous formations at Brodick church dip in a southerly direction, resting immediately above the old red

* These phenomena can perhaps be best examined on a small scale, where a number of the turns of the stream come under the eye at a glance. A beautiful example of it on a large scale occurs on the Clyde, at that part of Glasgow Green called "The Flesher's Haugh."

sandstone, which forms part of Glen Shirrag, and divides it from Glen Rosa; and the position of the high bed of limestone, on the north side of the Windmill Hill, which preserves the same dip, and forms part of the same bed as the limestone at the church, clearly shows, that part of the formations being destroyed, they are now separated from the subjacent old red sandstone strata by a wide denudation, which has hollowed out Glen Shirrag, and formed part of the alluvial plain of Brodick, in the manner described in figure 6.

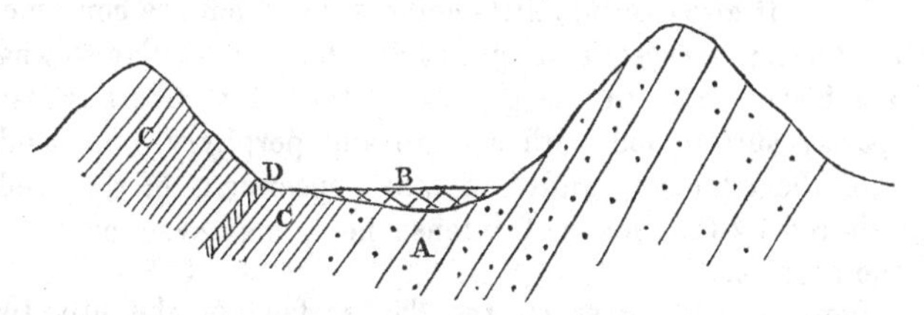

Fig. 6. Denudation of Glen Shirrag.—A the Old Red Sandstone on the north of Glen Shirrag; B the Alluvial hollow of Glen Shirrag; C the Carboniferous Formations with Limestone strata D, apparently unconnected with the subjacent Old Red Sandstone.

The north side of Glen Shirrag is therefore composed of old red sandstone; and the south side of the carboniferous series, with the exception of the steep summit of the Windmill Hill, which is composed of a claystone porphyry, an igneous formation, overlying the sandstone. The base of this porphyry is of a pale yellow or cream colour, and encloses numerous small crystals of quartz, and of common and glassy felspar of various sizes. The quartz varies in shade from white to brown, and not unfrequently assumes the form of small six-sided pyramids. One or two trap dykes traverse the porphyry on the south side of the hill. Near the south-west extremity of the peak, at the top of a small ravine, a bed of beautiful shale, nearly vertical, has been laid bare by the winter torrents, and is capped on all sides by the igneous formations which have here overflowed its surface. There can be little doubt that the shale belongs to the carboniferous series, which, it thus appears, composes the whole of the Windmill Hill.

The isolated position of this porphyry, resting, as it does, on the very summit of the hill, forms an additional proof of the great extent of the denudation of the surrounding country, and consequent alteration in its physical appearance; for, being an igneous rock, it was undoubtedly, when protruded, in a state of fusion: it is therefore very improbable that it could have rested on the very summit of the hill, when it reached the surface, without also flowing down its sides. Either, then, the north side of the hill was of such a conformation when the eruption took place, as to prevent this overflow, or the porphyry farther down the hill has disappeared. In either case, the denudation has been very great.

A little to the south-west of the Windmill Hill, is a mass of fine grained granite, (Section No. 2,) forming part of the hills which surround Glen Dubh.* The author first observed this remarkable circumstance in the summer of 1837,† before which it had not been noticed. By reference to the map, it will be seen that it is completely separated from the great central mass of granite to the north, by strata many thousand feet in thickness, consisting of the clayslate, the old red sandstone, and the carboniferous series. It is composed of the usual ingredients, but the proportion of reddish felspar far exceeds that of the quartz and mica. It also exhibits little hollows, in which are contained small but well developed crystals of quartz and felspar, similar to what are found on a larger scale in the granite of the interior. It is associated, and partly surrounded with syenite and porphyry, and sends forth veins into the adjacent sandstone, while specimens of the sandstone, much altered by the effects of intense heat, may even be found inclosed in the granite.

From the depth of the moss on the flat surface of the hill, the boundaries of this granite have not yet been determined, but by a little labour, properly bestowed, this object might, doubtless, be accomplished.

The existence of this fine granite, in such a locality, is of the utmost importance, and the facts connected with it ought to be particularly borne in mind, as they are intimately related to the proper development of the theory of the arrangement, and relative ages of the various formations, as attempted to be explained in the concluding chapter.

The rest of the district surrounding Glen Dubh, is principally composed of a very beautiful syenite, which differs only from the granite in the substitution for mica of hornblende, of which mineral it here contains about a third. It is utterly impossible to define the boundaries which separate the rocks of all these various igneous formations. Probably no such boundary exists, for rocks of this description frequently pass by degrees into each other, which indeed possibly takes place here. Both the granite and syenite have burst through, and now overlie, the new red sandstone conglomerate, which, in the bold cliffs of Craig na Iolaire, (or the Eagle Rock,) and Craig na Fitheach, (or the Corbie's Rock) rises from the dark hollow of Glen Dubh.

A little to the South West of Glen Dubh, where the ground begins to decline to the south, a mass of quartz rock, (Section No. 2,) is laid bare in the hollows of the moss. This rock has been altered from the common sandstone to its present form, by contact with syenite. It is of a light grey colour, and contains nodules of pure white quartz, which were

* The upper part of Glen Cloy is called Glen Dubh, or the Black Glen. There are two glens in Arran of this name; one forming part of Glen Cloy, the other branching off North Sannox.

† I mentioned the fact of the existence of this granite to Professor Nichol at the time, but the public are indebted for the first notice of it to M. Necker, who discovered it in 1839, and named the district Ploverfield.

probably, before it was altered by heat, quartz pebbles imbedded in the conglomerate, and which now form the greater proportion of the agregated pebbles of the neighbouring new red sandstone formations. To the east of the cliffs which overhang the Glen, rise the Sheeans, three round eminences, forming the extreme southern boundary of Glen Cloy. They are all composed of different varieties of trap, which, from thence, extends along the hills towards Lamlash.

Descending from the Sheeans to the east, a well known pitchstone dyke, about 30 feet wide, may be seen crossing the old Lamlash road, and passing through the new red sandstone, which is not in any way altered at the point of contact. It is of a dark leek-green colour, and contains a few small crystals of felspar. When splintered it exhibits a lamellar structure, more particulary after the process of decomposition has begun, when the shining gloss of the fresh fracture gives place to a thin white coating on the exposed surface. To avoid repetition, it may here be mentioned, that there is a similar dyke on the south side of Brodick Wood at the entrance to Glen Cloy, which also contains small crystals of felspar and a few of quartz. The largest and most important of all the pitchstone veins occurs on the Corriegills shore. It has been noticed by Dr. M'Culloch, whose minute description is here transcribed. "It resembles so strongly a mass of prismatic trap, as to be often overlooked in walking along the shore, even by those who have been directed to the spot. From its horizontal position, it has been by some called a bed, but like traps in similar cases, it may with propriety be considered a horizontal vein. The visible face is rudely prismatic, and about 12 feet thick; extending for about 200 yards, and terminating abruptly at each end. It is apparently conformable to the sandstone on which it lies; but whether rigidly so, cannot be discovered, as the faces of the cliffs are obscured, both by the mouldering of the rocks, and by dispersed patches of vegetation. Like the sandstone, it reclines to the south-west at an angle of about 30 degrees. Its texture is tolerably uniform throughout, being most commonly also lamellar, and it is of a dark, or bottle-green colour. It is not accompanied by any visible disturbance of the adjoining sandstone, nor is there any apparent change in either rock at the places of contact; except that, as happens so frequently among the trap rocks, the vein decomposes to a certain depth near the junction."

Between the pitchstone vein of the old Lamlash road, and the hills above Corriegill, there is nothing of sufficient geological interest to warrant a particular account of its formations.

The new red sandstone which succeeds the carboniferous series in the order of superposition, extends from Glen Dubh, along the hollow and the south side of Glen Cloy. (Section No. 1.)

Glen Cloy is therefore partly bounded on the north by the sandstone of the carboniferous series, which forms the principal part of the Windmill Hill, and on the south by the new red sandstone. From the entrance to Glen Cloy, the same formation, often in a conglomerate form, occupies the lower hills, the rocky beach, and the high cliffs which overhang the sea shore around the coast of Corriegills. It is at many places intersected by a few porphyritic, and by many trap dykes, principally greenstone, which may be seen for a short distance, till they are lost in the soil traversing the strata, in the form of veins, being generally cut sharply off at the surface of the penetrated rock. A particular account of these will be found in a succeeding chapter, since to mark each dyke as it occurs, would occasion a constant interruption in the description of the more extensive formations.

The higher part of the hills along this coast are almost entirely composed of dark coloured syenite, in which the hornblende much predominates in quantity over the quartz and felspar. It rests on the new red sandstone, which it has burst through and overflowed to a great thickness. This is beautifully developed in the cliffs between Dunfion and Clachland Point, where the horizontal sandstone strata forming the lower part of the precipice are surmounted by a deep covering of syenite, the exposed surface of which assumes a rudely columnar form, and gradually tapers away till it disappears at Clachland Point. At Clachland Point several greenstone dykes penetrate the strata of sandstone, which is much altered, being highly indurated at and near the point of contact.

The conspicuous hill called Dun Dubh, is not of the same formation as the surrounding rocks, but is composed of a claystone porphyry, somewhat similar to that of the Windmill Hill. It also overlies the new red sandstone, and near its summit is arranged in the form of clusters of regular four and six sided columns, dipping in various directions and at various angles, but principally about 60° N.E.

Having thus concluded his examination of the principal geological phenomena which surround Brodick Bay, the geologist, as he returns through the alluvial plain of Brodick, will remark that the soil is principally composed of the debris of the neighbouring hills, washed down by the ceaseless torrents; and on examining the sand and pebbles on the curved and sloping beach, he will observe that disintegrated granite enters most largely into its composition, intermingling with the waste of the clay slate, the sandstone, and with quartz and other pebbles, from the surrounding conglomerates. He will also observe the gradual manner in which the land is still gaining on the sea, by the formation of sand banks at the mouths of the streams, and thus be led to speculate on the remote period, when, possibly, the waves of the sea dashed far into the peaceful glens, now covered with vegetation. It is well known that the Duchess Ann built a ship of war of Arran oak, and presented it to the British government. This ship was launched from what is now an enclosed field, below the little garden at the north end of the village of Brodick, and which is now at least two hundred yards from water of sufficient depth for such a purpose. There is also a tradition of an anchor having been found buried in the soil, somewhere in Glen Rosa. Those, however, who are acquainted with the story of the discovery of the anchor on the top of Table Mountain, might perhaps regard such evidence in rather a doubtful light.

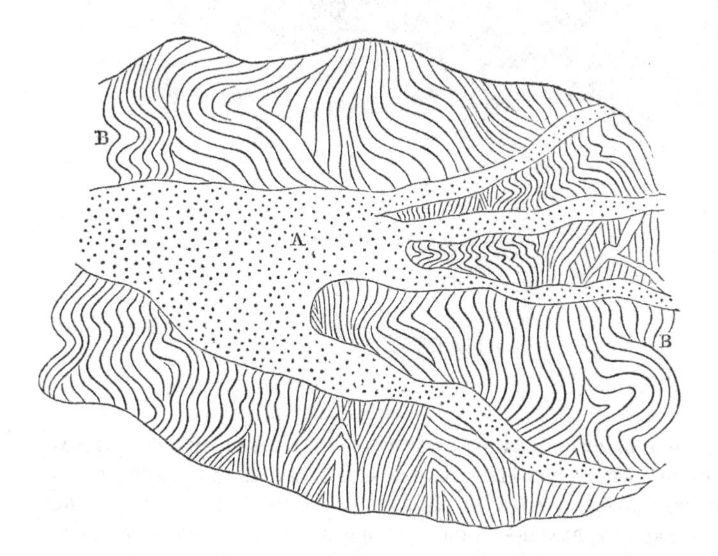

Fig. 7. Granite veins penetrating slate, east side of Glen Rosa. A Granite. B Slate.

Glen Rosa from Brodick shore.

CHAPTER II.

FROM THE MILL DAM TO THE WHITE WATER—SYMPTOMS OF THE APPROXIMATION OF GRANITE AND SLATE—JUNCTION OF GRANITE AND SLATE AT THE WHITE WATER—LIMESTONE AT MAOLDON—ANCIENT SEA CLIFF ON THE COAST—ANCIENT BEACH—CARBONIFEROUS BEDS AT CORRIE—IGNEOUS ROCKS IN CONNEXION WITH THEM—OLD RED SANDSTONE—WATER-WORN CAVES IN THE CLIFF—CONICAL BOULDERS ON THE SHORE—SULPHATE OF BARYTES—ITS MANUFACTURE—DISAPPEARANCE OF SLATE NEAR THE GRANITE OF GLEN SANNOX ACCOUNTED FOR—GLEN SANNOX—FROM THE HILLS WHICH DIVIDE IT FROM GLEN ROSA.

HAVING concluded the survey of the environs of Brodick Bay, the geologist will now bend his steps round the north-east coast, which, from its variety, and the importance of its formations, forms by far the most interesting geological district of Arran.

On the coast between Brodick Bay and Port-na-Claoch, the rocks, which are of the carboniferous series, with the exception of a few intersecting greenstone dykes, afterwards to be described, exhibit nothing else of peculiar interest. There are, however, higher on the hills, many phenomena well worthy of notice.

Part of the boundaries of the carboniferous and other formations on the hills around the base of Goatfell, are obscured by an extensive moss, and their exact extent being indefinable, the moss is marked in the map as alluvium.* The same difficulty does not occur regarding the junction of the slate with the granite, for although a patch of moss also interferes to cover the visible junction, and though there are no streams of water in which the strata can be examined between the mill dam and Maoldon, yet, from various circumstances, the line of demarkation can be pretty accurately determined. It is, of course, never possible to trace, step by step, the actual contact of the granite and slate, as this junction is often covered with soil, alluvium, and moss, without any intervening streams for considerable distances. Still, when they are not positively seen to join, an experienced eye can often judge by appearances that such junctions are not far distant. In many places where the general face of the country consists of extensive marshes or mosses, little slaty knolls and rising grounds present themselves here and there, frequently almost bare of vegetation, and skirting round the base of the more elevated and peaked part of the mountains. From the contortions of the strata of the schist or slate, the increased number of quartz veins, the induration of the rock, and other symptoms, it is generally possible to tell if it approach the granite, though, as will afterwards be seen, some of these appearances not unfrequently occur at comparatively great distances from such junctions. Such a chain of small rising grounds extends round the granite, from Glen Rosa to the mill dam, and from thence to the neighbourhood of Maoldon; and it is by their aid that we are here enabled to trace the boundary of the slate. It should always be borne in mind by the geologist, that no appearance on the surface of a country is too unimportant for observation, as it is often not till after much reflection, that he becomes aware of the importance of what at first sight appeared totally insignificant. In the concluding chapter it will be seen that these observations, besides indicating the approximate bounds of the slate, partly solve another geological problem.

As might be expected, in the first little stream north from Maoldon, the actual junction of the granite and slate is visible, and from the alternations of these rocks in the bed of the stream, which here is not more than two feet broad, it is evident that the slate is here interpenetrated by granite veins. The slate is much contorted, which, as before stated, is invariably the case, unless, as in a few instances, the marks of stratification are entirely obliterated.

One of the most interesting and instructive junctions in Arran is in the next stream to the north, at the foot of a high and very conspicuous cascade called the White Water, which, as seen from the road, presents

* As it is often impossible accurately to determine what formations underlie these mosses, it is coloured in the map as alluvium, and this principle is always adopted whenever the existence of mosses, soil, or accumulation of debris involves a geological question.

one unbroken line of white and sparkling foam, as it dashes down the mountain side. The actual junction takes place at the foot of the fall, and is somewhat difficult of access. Unless closely examined, it might be supposed that the secondary strata rest directly on the granite; but a closer inspection shows, that a very narrow strip of slate intervenes between the granite and sandstone. There is no distinct line of boundary between the sandstone and slate, the latter gradually merging into a dark coloured sandstone, which again changes into a coarse conglomerate. It is also worthy of observation, that the line of demarkation of the granite and slate, (two rocks, it will be remembered, formed by different agents, viz., fire and water,) is in some places destroyed, the marks of stratification wholly disappearing, and the slate assuming a light bluish colour, streaked with white, and in some places exhibiting crystals of felspar, as if it had been partially fused by contact with the intensely heated granite, the lighter coloured materials of which are intermingled with the darker substance of the slate. As the granite approaches the slate it becomes of a finer texture, being, indeed, when close to the slate, altered into a compact felspar, which, again, is also partially affected by the proximity of the slate, being slightly tinged with blue. In this state it is much more brittle than the coarse grained granite into which it merges, a little removed from the junction.* Though not always so fine, or so extensively developed as in this instance, it will generally be found, with a few exceptions, that a narrow strip of compact granite intervenes between the crystalline granite and the slate; and this, as was formerly remarked in reference to the fine veins of the two granites, is probably owing to its coming in contact with a cooling substance (the slate,) in which case, the crystallisation proceeding more rapidly, the crystals were necessarily smaller than when the melted matter was farther removed from the cold. For some distance north of this stream, the stratified rocks, in the neighbourhood of the granite, are lost in the moss.

A little S. E. from this junction, on the north face of the round hill called Maoldon, there is a great upcast of lime, (Section No. 1,) similar in every respect to that at Corrie, which is afterwards particularly described. They are both of the same colour and quality; they contain the same shells, and the beds of limestone and shale alternate with each other, in the same manner.

It will have been observed, that an ancient sea cliff overhangs the narrow plain intervening between the sea and the ascent of the hill to the north of Brodick. Between this cliff and the road, in what is now at many places ploughed fields, numerous recent shells, often in a perfect state of preservation, are mingled with the soil. With a few exceptions, they are almost entirely of the same species as those now existing on the modern beach. The following were picked at random

* It is necessary to climb up part of the waterfall, with the risk, or rather the certainty of being thoroughly wetted, to get a full view of all these phenomena.

from a field about a mile south of Corrie :—Cardium Lævigatum, Patela Vulgaris, Lucina Radula, Purpura Lapillus, Turbo Littoreus, Turbo Rudis, Trochus Cinerarius, Trochus Magus, Trochus Crassus, Nerita Littoralus, Venerupis Decussata, Venerupis Palestra, Mytilus Edulis, Buccinum Undatum, Venus Fasciata, Cytherea Exoleta, Terebra Reticulata, Rissoa Calathisca, Rissoa Semicostata.* The presence of these shells in such a locality, sufficiently indicates, that what is now cultivated ground, was formerly the sea shore, which must, therefore, have been elevated to its present position above the tidal level, by subsequent upheaving agencies.

The various beds in connection with the formations at Corrie, begin where the carboniferous series overlies the old red sandstone, at the old march of Achab Farm, where the geologist ought to begin his observations on the different strata, as they occur in the ascending series. Their upturned edges are exposed on the shore, where they can be most effectually examined.

The underlying strata of old red sandstone here assume the form of a rather fine conglomerate ; but before passing finally into the coal measures, we find the same pebbles of which the conglomerate is composed, imbedded in lime, and forming a calcareous conglomerate, (dip 15° S. E.) after which succeed the ordinary beds of limes, shales, and sandstones. There is, therefore, no sudden change from the one set of strata to the other. The transition is perfectly gradual, and in perfect harmony with those theories resulting from the researches of modern geologists, which all tend to demonstrate that the old hypothesis of vast and sudden convulsions, revolutionising at stated epochs the whole surface of the globe, have no place in the economy of nature, where every thing is, and ever has been, regulated by laws, whose grand characteristic is the principle of a slow but certain progression.

A bed of grey sandstone rests on the calcareous conglomerate, and is again succeeded by concretionary limestone or cornstone, consisting of nodules of red lime imbedded in a soft argillaceous shale. This is followed by a few feet of grey lime (dip 29° S. W. by S.) subjacent to a thick bed of sandstone, extending about forty-two feet along the shore, and which is again succeeded by narrow beds, consisting of alternating strata of sandstone and shale. To these succeed a great trap dyke, extending about three hundred and twenty paces along the shore, and which most probably has burst through and overflowed strata similar to those already described. This dyke presents a considerable variety of forms, being in some places a compact basalt, containing a great proportion of hornblende, and at others assuming an amygdaloidal structure, the porous cavities of which are filled with white and pale yellow crystals

* Of these shells, Venerupis Decussata is rare in the west of Scotland. The Rissoa Calathisca, though found on the coast of Largs, is rarely seen on modern beaches. It is disputed by some naturalists whether or not the Trochus Crassus is ever now found on modern beaches in Scotland. It is, at all events, exceedingly rare.

of carbonate of lime. It often contains numerous crystallised veins, some of which are very minute, and others about ten inches broad. The small veins are principally composed of carbonate of lime, and sometimes of quartz; the larger ones of carbonate of lime, quartz, and steatite, which are sometimes united in one vein. When subjected to the action of the weather, the exposed surface peels off, as it decomposes, in thin layers, like the concentric coats of an onion, exhibiting large rounded masses on the surface of the rock, thus: (Figure 9.)

Figure 9. Decomposing surface of Trap at Corrie.

As all the igneous rocks of the trap family decompose in the same manner, this circumstance must therefore be dependent on some law connected with their peculiar conformation. For some distance beyond this formation, the beach is covered with gravel, at the south end of which is a stratum of red limestone, associated with a thin bed of grey marble, which is capable of receiving a tolerable polish. Another trap dyke here penetrates the strata, and extending in breadth for a few paces along the shore, is succeeded by a light grey sandstone, which occupies the beach from Corrie gate to the quarry. The quarry of mountain limestone is about twenty feet thick, and consists of twenty-two beds of lime of a blueish-grey colour, alternating with an equal number of beds of red shale, which vary from a sixteenth of an inch to about a foot in thickness, the whole being surmounted by a bed of magnesian limestone, four feet in thickness. Its dip varies from twenty-nine to forty degrees, S. E. by E., (Section No. 1), and it is extensively wrought for architectural and agricultural purposes, by means of caverns which are excavated from the upper edge in the direction of the dip.

Its fossils are very abundant, consisting principally of the Producta Scotica, Sperifer Striatus, Cardium Alaeforme, various Encrinites, Madriporites, and, though rarely, Ammonites.

The larger Producta are so abundant, as almost entirely to form the lower layer of some of the limestone beds, being partly imbedded in the separating beds of shale, in which they uniformly rest with the convex side of the valve downwards, indicating the tranquil condition of the bed of the sea where they lived and died. Several trap dykes penetrate the rock, the principal one of which is about fifty feet from the mouth of the cave, and may be five and a half feet broad. At the point of contact the lime is considerably altered, being indurated, harsher to the touch, and of a darker colour. Another, four and a half feet broad, is immediately beside the last, and on its west side the beds of limestone are raised three feet. About a hundred feet further west, is a third dyke penetrating the same strata. The dip is here 20° S. 20° E., while the inclination of the first and last dykes now mentioned, is in both 48° S.

70° W., the other being vertical. Their direction is nearly north and south. The following formations immediately overlie the limestone:—

1.	Variegated and common shales, containing a bed of hematite, three inches thick, ...	7 feet.
2.	White sandstone, ...	10 —
3.	Blue shale, ...	1 —
4.	White sandstone, ...	6 —
5.	Shale, containing hematite six inches thick, .	5 —

Similar alternations occur above these, which it is unnecessary to particularise.

In the white freestones associated with this formation, branches of Calamites and Stigmarias frequently occur; a specimen of the latter was found, in one instance, about a foot in diameter.

A short distance to the south of Corrie there is another upcast of limestone, (Section No. 1,) near the small cascade of Locherim Burn. It is part of the same bed, and contains fossils similar to those of Corrie.

It thus appears, that besides the upcast in Glen Shirrag, the limestone is thrown to the surface four times between Brodick and Corrie,—in the woods at the west side of the castle, at Maoldon, where it rises nearly a thousand feet above the level of the sea, at Locherim, and at Corrie. These upcasts, (Section No. 1,) according to Mr. Murchison and Professor Sedgwick, are the indications "of a series of dislocations, marked by corresponding advances of the granite from the central ridge, each of which seems to have produced an upcast of the limestone." It has indeed been supposed by some who have examined the district, that as all the beds of limestone between Corrie and Brodick dip towards the south, they are four distinct and separate beds, and, along with the other strata, are superimposed on each other agreeably to the dip of the anticlinal line. After carefully examining the district, it seems impossible to hold such an opinion, and there can be no doubt of the truth of that already quoted; the more so, because the limestone and shales of the different upcasts, present a perfect uniformity of colour and structure, and contain the same fossils, the Producta Scotica predominating in all of them. It will be afterwards seen that this is not the case, in the various beds of limestone to the north of the anticlinal line, between the Fallen Rocks and the Cock, where they are not only of entirely different colours and texture, but also contain different proportions and species of fossils.

It was formerly stated that the carboniferous series rested on the old red sandstone, a little to the north of Corrie. To this point it is now necessary to return. It will have already been remarked, that the carboniferous series between Corrie and Brodick reposes directly on the clay slate; but as it is also superimposed on the old red sandstone at Corrie, agreeably to the dip of the anticlinal line, (see Section No. 1,) it follows, that the latter formation, in regard to the relative antiquity of the stratified formation occupied during its deposition a period intervening between the formation of the clay slate and the rocks of the carboniferous series.

At this part of Arran, the old red sandstone (Section No. 1,) occupies comparatively a large extent of country, extending far inland, and stretching from near Corrie to the Fallen Rocks, a range of coast of about three miles in length.

It exhibits some varieties of structure, varying in fineness of texture from a compact red sandstone to a coarse conglomerate or pudding-stone, in which many of the imbedded pebbles are at some places, as in North Sannox Water, nearly a foot in diameter. In the conglomerate form, it is principally composed of water worn pebbles of slate and schist, greywacke slate, pieces of quartz, and a little jasper, imbedded in a dark-coloured sandy cement. These pebbles are generally well rounded, as if long subjected to attrition by the action of the water. The fragments of quartz, however, present an exception to this rule, being of two kinds; first, well rolled and polished pebbles, possibly originating in some distant and ancient mass of quartz, which has now disappeared; second, angular fragments which seem to have been imbedded in the conglomerate immediately after they were detached from the parent mass. These last probably proceeded from the larger veins of quartz (afterwards to be noticed) so numerous in the slates and schists, the softer materials in which they occur, forming part of the cement which binds the conglomerate together.

Another appearance of common occurrence in this formation can also be sufficiently well observed in the district between Corrie and Sannox. The rock, in its conglomerate form, frequently alternates with narrow beds of fine sandstone. These alternations have sometimes been supposed to indicate partial elevations and subsidences in the bed of the sea where the materials were deposited; but from their frequency, it seems much more probable that they ought to be attributed to the periodical increase and diminution of those currents or streams which bore the gravel to the ocean.

In the face of the ancient sea cliff which has already been noticed as extending along this coast, are a number of water-worn caves, the most strongly marked of which occurs in the face of a beautifully wooded cliff, immediately beyond the second gate south from Glen Sannox. (See the view in page 61.)

The existence of caves in such localities, is generally considered as an indication that they owe their formation to the action of the waves beating at some former period against the cliff. They may therefore be considered, in connexion with the shells which are found between them and the modern beach, as forming an additional proof of recent elevation.

On the south side of the same gate, another well marked phenomenon also points to the fact of a recent elevation. Between the road and the sea are two large boulders of granite, represented in page 41, which have rolled down from the neighbouring heights. These stones, which are now considerably above the tidal level, rest not on their broadest and most solid part, but on their apices, as if, while they were within high water mark, the action of the advancing and retiring waves had washed away the lower portion of the rock, and left them, when the coast was upheaved to its present height, in the position they now occupy. An approach to a regular series of these forms may be observed on the shore between Brodick and Glen Sannox: and though it cannot by any means be said to be an invariable rule, yet it cannot escape observation, that the tendency to assume this form is such, that, in a number of instances, the further the boulders are removed from ebb tide, the more do they assume the appearance of inverted cones.

A peculiar appearance, exhibited on the surface of the sandstone on the beach between Brodick and Glen Sannox, cannot fail to attract attention. It will have been observed, that a number of elevated ridges, varying from a few inches to six or eight feet in height, cross the sandstone on the shore, frequently intersecting each other, and generally exhibiting rough and serrated edges. In some instances they are doubtless caused by trap dykes hardening the rock to such a degree, that it resists the action of the elements longer than the unaltered and softer strata, which more rapidly yield to decay. But in the majority of cases they do not owe their origin to this cause. It is also evident that they are not prolongations of harder strata, enclosed between soft beds of sandstone, since they generally project above the surface, at a much higher angle than the prevailing dip of the strata, which, on a great part of this coast, is not of great elevation. Possibly they may be caused by the formation of fissures, which originated during some ancient disruption of the strata. Into these sand may have gradually settled, which, from causes it may not now be possible to ascertain, assumed a firmer texture than the original stratified deposits. Any one who wishes to determine this point, may do so by observing whether or not the line of stratification be prolonged through the ridges, where they enter the main body of sandstone. If it be not prolonged, then their existence may be accounted for in the manner shown above; but if the line be carried through the ridges, it is impossible that they could have originated in this manner, and some other cause must be sought for.

Pursuing his investigations along the shore, the geologist finds the path diverge into Glen Sannox, where several appearances of much interest occur. Following the course of the stream, the same alternations of sandstone and conglomerate are visible. Near the entrance to the Glen these strata contain a vein of sulphate of barytes, which crosses the bed of the stream, N. E. by E., dipping into the rock at an angle of 51° N. W. by N. The same vein rises to the surface on the hill to the S. of Glen Sannox, in the bed of a small torrent originating in Corrie na Chiodh. This mineral, when pure, is of a white colour, and opaque in the mass; but much of it is more or less impure, being slightly tinged with yellow, from admixture with oxide of iron. It is of great weight, its specific gravity being 4.4, and it principally occurs in a massive form, though imperfect crystals are occasionally found on its surface, and in fissures.

It is now extensively wrought for economical purposes, and a mill has lately been erected in Glen Sannox, where it is manufactured for the market. It may not be uninteresting to describe the process of manufacture. When taken from the quarry, it is conveyed on a wooden railway to the mill, where it is first crushed under two upright revolving rollers, which are set in motion by a water wheel, worked by the water of the burn. This breaks it into a coarse powder. It is then put into a grinding mill, consisting of a large round cistern about fourteen feet in diameter, the bottom of which is composed of slabs of granite, about fifteen inches thick, coarse grained, and as hard as possible. These are strongly bound together, and neatly fitted so as to contain water. In the centre revolves an upright shaft, from which radiate several cross arms. To these are attached by strong chains, large masses of granite, flattened on the bottom, and weighing about ten hundredweight each. At a certain height in the cistern there is an overflow pipe. This cistern being filled with water, the coarse powder is put into it. The shaft is then made to revolve, and the cross arms attached to it drag round the heavy stones, which now reduce the barytes to a fine powder. The troubled motion of the water raises the finer portion of the powder, and it escapes with the superfluous water by the overflow pipe, into another cistern. Here as much sulphuric acid is added to the water as makes it strongly acid. This mixture dissolves the iron and other metallic impurities. The acid water is then run off, and clean water added to wash the barytes. This is again run off, and the barytes, when about the consistence of paste, is put on a long flat stove, the top of which is of stone, surrounded by an edging three or four inches high. As it hardens, it is cut into oblong cakes, and when almost firm, it is stored on open rack-work shelves in another stove, there to be dried thoroughly.

It sometimes undergoes a colouring process, in which case it is dyed before it is put into the first stove, thus: If to be dyed yellow, it is put into water, and well mixed along with a small quantity of nitrate of lead. The nitrate of lead dissolves. Into this is poured a quantity of dissolved chromate of potash; chromic acid leaves the potash and unites with the lead, forming the yellow salt chromate of lead, which is insoluble in water, and consequently falls to the bottom. Besides the slight admixture, a coating of the chromate of lead is thus formed round each particle of the barytes. It is then dried in the manner already described.

To the westward of the sulphate of barytes, the old red sandstone continues to form the bed of the stream; and as it approaches the granite, some of its lowest beds assume the character of a very compact greywacke slate. This formation suddenly disappears, where it might be expected to be found resting on the ordinary slate, and a few yards further up, the water flows over solid masses of coarse-grained granite. The intervening space between the old red sandstone and the granite, is completely overwhelmed with deep beds of gravel and massive granitic blocks, thus rendering it impossible to determine positively on the spot, whether or not the covered intervening space be occupied by the expected strata of slate. By an examination, however, of the neighbouring hills, an approximation to the truth may be attained. And in entering on this investigation, it is necessary to include some remarks on another disappearance of the slate somewhat further south.

A reference to the map will show that the granite is entirely surrounded by the clay slate and schistose series, with the exception of two places where the circle is broken,—first in the bottom of Glen Sannox, and again in a small district of moorland on the low hills between Maoldon and Corrie. At the foot of the granitic portion of Suithi Fheargus,* a hill forming the east end of the range on the north side of Glen Sannox, a junction of the granite and slate is visible, the granite penetrating the slate in narrow veins. The strip of slate which here intervenes between the granite and the old red sandstone is very narrow, not more than fifty yards wide ; and, descending the hill towards Glen Sannox, it gradually becomes still more confined, and is finally lost in the moss. As already stated, it does not reappear in the bed of Glen Sannox burn, where, if it exist at all, it is most naturally looked for. It is however extremely probable that a narrow strip of slate exists in this locality, concealed under the obstructions already referred to. A similar instance, without these obstructions, occurs at the bottom of the cascade between Maoldon and Corrie, where the junction of granite and slate takes place. In the description of this junction, it was observed that the strip of slate intervening between the granite and sandstone, is only a few yards in width, after which it merges into a gritty sandstone, which again gradually changes into a coarse conglomerate. But had the junction been a few yards further from the cascade, in the more level bed of the stream, it also would have been obscured by gravel and boulders, and to all appearance there would have been no intervening rock between the granite and sandstone, thus inducing the same question that arises from the disappearance of the slate at Sannox. What increases the improbability of a junction of granite and sandstone in Glen Sannox is this, that on the hills on the south side of the Glen, at the foot of the steep conical mountain called Ciodh Na Oigh,† a narrow strip of slate again appears, similar to that at the foot of Suithi Fheargus, and which is therefore probably continued across the Glen, beneath the obscurations, to the base of that hill. The slate can be

* Suithi Fheargus, or Fergus' Seat. There is a tradition in Arran, that, once on a time, when Fergus the First made a survey (not geological) of his kingdom, in the course of his wanderings he visited Arran. To obtain a view of the island and the frith, the monarch and his attendants climbed this lofty hill, where, to refresh themselves after the fatigue, they sat down to dine. Hence the name.

† Or the maiden's breast.

traced in occasional junction for about a mile to the south of Ciodh Na Oigh, when it is again lost in an extensive moss, which covers the flat summits of the low hills to the west of Corrie. In all likelihood the slate is also continued beneath this moss: in proof of which the granite, which is here coarse and crystalline, if carefully observed, will be found at some spots to assume that fine compact texture and blueish tint which it so often assumes when in contact with, or approximating to the slate. It thus seems probable that the granite is wholly surrounded by slate, though it is impossible to prove to a demonstration that such is the fact.

Above this point, the remainder of Glen Sannox is entirely composed of coarse-grained granite. In penetrating into the Glen, however, it may be observed, that the large rounded boulders, profusely scattered over its surface, change their character, many of them being of the fine-grained granite, already noticed as forming the interior of the granitic district. This is accounted for by the circumstance that an immense vein of fine granite, crosses the north ridge between Suithi Fheargus and Ceum Na Cailleach, from which the boulders have gradually been detached and rolled into the Glen. As yet no trap dykes have been observed in the hollow of Glen Sannox, but on the northern range, there is a remarkable fissure called Ceum Na Cailleach,* forming a deep indentation on the summit of the mountain, from whence a narrow gully descends into the Glen. This seems to have been formerly entirely filled with a trap dyke, the decomposition of which has left the more durable granite in its present conspicuous form; and the direction in which the dyke runs renders it probable, that along with three other appearances of the same porphyritic trap on the west side of Glen Rosa, it forms one great dyke, now only exposed on the surface at these four points. There are several trap dykes in and around Ciormhor, a high conical hill, forming alike the upper extremities of Glen Sannox and Glen Rosa. One of these, of a very singular construction, crosses the ridge which divides Glen Rosa and Glen Sannox on the south side of Ciormhor. This remarkable dyke includes five distinct bands, which are composed of three descriptions of rock. The middle is of green pitchstone. On each side is a band of claystone porphyry, two feet broad, each of which is again bounded by bands of trap, eighteen inches broad. This dyke forms a deep ravine, which reaches to the summit of the mountain, when the claystone disappears, the other materials composing the dyke being somewhat farther prolonged. The trap is partly decomposed in concentric layers, exhibiting on its surface bullet-shaped boulders of various sizes.

During the course of his investigations, as the geologist pursues his lonely path up the glen, he is ever induced to pause and gaze around him, in silent admiration of the wild and desolate scene. No sound of

* Or the carlin's step.

man strikes his ear, save the click of his own hammer reverberating from rock to rock, and no sign of life is heard or seen, but the bleating of the moorland sheep, the wild shriek of the eagles soaring from their eyrie in Ciormhor, or, if it so fortunately happen, he may chance to spy in the distance, the stately forms of the red deer pacing silently up the mountain side. And ere he turn his homeward steps into the long vista of Glen Rosa, he will linger on the rocky ridge between the glens, and there, amid the lofty precipices, and surrounded by the grey peaks of the solemn hills, he will reflect, that with all their appearance of majesty and power, day by day they are slowly crumbling into dust; and even now the landscape on which he mutely gazes, is imperceptibly yielding to the never dying principle of change; and the time will surely come, when, with all its varied features, it shall have passed away, and left no trace behind.

Glen Sannox.

Loch Ranza.

CHAPTER III.

FROM GLEN SANNOX TO LOCH RANZA—OLD RED SANDSTONE—ANTICLINAL AXIS—FALLEN ROCKS—CALCAREOUS CONGLOMERATE—ALTERNATIONS OF THE CARBONIFEROUS SERIES FROM THENCE TO THE COAL PITS—COAL AND ASSOCIATED BEDS, ETC.—NEW RED SANDSTONE—THE SCRIDEN—ITS ORIGIN—UNCONFORMABLE STRATA AT ALT BEITHE—LOCH RANZA—ALLUVIUM—INTERIOR BETWEEN GLEN SANNOX AND LOCH RANZA, INCLUDING GLEN CHALMADAEL AND THE JUNCTIONS OF TOIRNANEIDNOIN AND THE GREAT JUNCTION IN GLEN DUBH.

A GEOLOGICAL survey having thus been completed, including all the principal features of the country around Brodick Bay, and also of the coast and interior from Brodick to Glen Sannox, the geologist will again begin where he diverged into the interior; and as the way is long and rough, and he has much work before him, let him make an early start, to continue his investigations along the coast from Glen Sannox to Loch Ranza.

Between South and North Sannox waters, the old red sandstone conglomerate runs in a high mural escarpment called the blue rock, the lines of stratification of which dip at a low angle towards the south. Immediately to the north of North Sannox water, a low cliff of the same formation runs along the shore; and this is a position of great geological importance, for at this point is placed the anticlinal axis. (Section No. 1.)

It will have been observed that the strata already examined have two dips, one of which is caused by the inclination of the rocks against the central granite, the other consisting of an inclination by which the coast strata uniformly dip towards the south. And further, the angle of this last inclination, when not interrupted by faults, (Section No. 1.) uniformly decreases the more the strata approach to this point, till, a little to the north of North Sannox, the lines of stratification become horizontal, and finally, a little further north, begin to dip in the opposite direction.

The point, then, between these opposite inclinations is the *anticlinal axis.*

Proceeding to the north, the strata are for some distance nearly horizontal. About two miles from Sannox, the eye of the traveller is suddenly arrested by a scene as imposing as it is unexpected, known by the name of the Fallen Rocks. An immense cliff of old red sandstone conglomerate, which overhung the brow of the hill, seems suddenly to have given way, and the entire slope is covered with huge irregular masses of rock, hurled from above, in the wildest and most tumultuous confusion, and which now strew the face of the hill from its summit to the sea. It is on the beach immediately below, that this view is seen in its most impressive aspect. There, as the spectator surveys the massy fragments, he can scarcely but fancy that the crash and the roar of the descending avalanche has but newly ceased, and the impending ruins seem even now about to descend, in wild tumult to the shore.*

A little farther north the old red sandstone is succeeded by the beds of the carboniferous series (Section No. 1.) It is difficult to denote the exact line of demarcation where the two formations meet, since though the mineral character of the intermediate beds is perhaps more allied to the succeeding undoubted carboniferous deposits, yet, as they are destitute of fossils, it is impossible accurately to determine to which series they belong; and they therefore may properly be considered as denoting a transition from the one set of strata to the other, in this instance, thus again indicating the principle of gradual progression. Here, as at Corrie, these beds do not pass suddenly into the carboniferous series, but are succeeded by intermediate beds of hard calcareous conglomerate, containing many quartz pebbles, associated with beds of highly indurated sandstone of a light red tinge, alternating with masses of indurated shale of a red, grey, and greenish-grey colour. Succeeding these formations are the regular beds of the carboniferous series, and their edges being invariably turned upwards throughout all this district, all their

* It is not known when this fall occurred, but there is a tradition that the noise of the descent was heard in the Island of Bute.

alternations can be accurately examined, except where obscured by shingle, thus affording an excellent opportunity of examining the carboniferous series. These strata alternate in great variety to the northern new red sandstone. But to give a minute account of each stratum would be both a useless and a toilsome task, and no practical benefit could possibly result from it. It is sufficient to describe the most interesting points, which sufficiently denote the nature of the formations, and the manner of their alternations, leaving it to the geological learner to trace out the analogous strata, as he walks along the coast.*

The beds succeeding those already described consist of various shales alternating with very hard light coloured sandstone, which are again followed by thin beds of shale with obscure impressions of plants, and by beds of sandstone. Overlying these in Lagantuin Bay, are beds of highly indurated shale, much altered by contact with a series of trap dykes, (Section No. 1,) which now principally occupy the coast for about a quarter of a mile, (Section No. 1,) during which space, these dykes present all the appearance of regular beds having the same range and dip as the superior and inferior strata of the coast Section.

They are as follows:—

1. Various amygdaloids abounding in crystals of carbonate of lime, and having some of the cells filled with finely crystallized green earth; others with a darks teatitic substance, and more rarely filled with zeolite.
2. Claystone with veins of green earth.
3. Light-coloured and earthy varieties of porphyritic greenstone.
4. Black porphyritic greenstone.
5. Dark porphyritic trap with an earthy base, containing crystals of diallage metalloïde.

These are succeeded by beds of white sandstone dipping 45° and 50°, N. W. W. The remainder of the shore of Lagantuin Bay is covered with small boulders, but towards the northern point the strata consist of alternations of white sandstone and red shale somewhat indurated, the latter being sometimes associated with red oxide of iron. In the ascending order these are succeeded by a long series of beds of white sandstone and grits, which, near Millstone point, are much dislocated and piled on each other in large blocks of a prismatic form. From thence, approaching the coal pits, the alternations of the principal strata are as follows:—

1. Blue shale with thin beds of calciferous sandstone and limestone, dip 34° N.E. by E.	about 25 feet along the shore
2. White sandstone, ...	2700 feet.

* This coast was formerly surveyed by the Rev. A. Sedgwick and Mr. Murchison, and to their admirable paper (the accuracy of which I have verified,) published in the Transactions of the Geological Society in 1819, I am indebted for such of the following observations as escaped my notice, or which I did not record during my survey.

3. Black mountain limestone (Section No. 1,) in thin beds, alternating with beds of black shale and sandstone, dip 33° N.E. by N. } about 120 feet.

   These contain many fossils, the principal of which are Producta Scotica, Producta Gigantica, a spiral univalve resembling Rostellaria, spines of Echinites, Caryophyllia, Encrinites, and rarely an Orthoceras.

   As at Corrie, these Productæ uniformly lie with the convex side of the valve downwards, again indicating the perfect tranquillity which prevailed in the bed of the ocean during the deposition of the strata. The larger varieties are frequently imbedded in the soft black shale, and can therefore be procured entire without difficulty; but the others being enclosed in harder substances, cannot easily be detached from the mass in which they are contained.

4. A series of beds of white sandstone alternating with black and coloured shales, with and without fossils, which are again succeeded by } about 2000 feet.

5. The strata of shale and sandstone immediately containing the coal (Section No. 1.) The coal is said to have consisted of three beds, but from the ruined and imperfect conditions of the workings which have partly fallen in, two can only be traced at present. The south bed rests immediately on a black bituminous shale, several feet in thickness, and containing numerous impressions of reeds and what appears to be flattened arborescent ferns. When newly split these stems are seen to be formed of an extremely brilliant coal, imbedded in black arenaceo argillaceous shale. It is very brittle, and apt to shiver and split into fragments when exposed to the smallest violence. The principal seam is said to have been about fourteen feet in thickness, but if we may judge from present appearances, it does not exceed three or four feet. It was formerly wrought from the upper edge downwards towards the sea in the direction of the dip of the strata, the roof being composed of sandstone and the bottom of the bituminous shale already noticed. This roof being partly fallen in, and the sea invading the seam at every tide, it is now extremely difficult to procure specimens of the coal. It is of that kind commonly known by the name of blind coal, which is almost destitute of bitumen. This description of coal contains more than the ordinary amount of carbon, and when ignited, produces an intense heat emitting neither flame nor smoke. When fractured, its layers exhibit a brilliantly polished surface.

Beyond the bed of sandstone forming the roof of the seam, are similar alternations of fossiliferous shale, coal, and sandstone. Several other pits have been sunk, now filled with water, and a gallery cut into the side of the hill where the coal and other strata must also extend, as the beds of limestone on each side of it, may both be traced to a considerable elevation on the side of the hill. The rock in which this gallery is cut, presents the curious though not uncommon phenomenon of curved strata, bending over the entrance to the cave in the form of a gothic arch. The same appearance may be seen in a less degree, in a bed of blue shale on the beach a little farther to the north. This curvature may either have been caused by an upheaving force acting on the middle of the strata, a resisting pressure being exerted at the two extremities, or it may have been formed by a similar cause ere the consolidation of the strata. Considering the high angle (45°) at which the seams immediately dip into the bed of the sea, and the manner in which they are partly intersected by trap dykes, it is matter of astonishment that they should ever have been deemed worth the risk and expense of working. The coal was wrought principally or altogether for the purpose of supplying with fuel the salt-pans, the ruins of which stand on a bank in the immediate vicinity.

A peculiar air of solitude reigns around this quiet place. The shore is rocky and precipitous, and the deep clear water almost approaches the grassy bank, along which an uncertain track marks the footsteps of the solitary traveller. The mines are now deserted, save by the destructive otter, which here finds a secure retreat in the sea-washed caves. The grey ruins still stand on the unfrequented shore, and though uninteresting in themselves, in this tranquil spot, lend an additional charm to the still and lonely landscape.

For a short distance on the beach, the strata are now obscured by shingle. Beyond this, alternations of white sandstone and shale occur, similar to those already described. The surface of the sandstone is frequently honey-combed, and many of the variegated shales are exceedingly beautiful, exhibiting a great variety of spotted colours, shaded into each other in endless change. This range of strata also contains three beds of red oxide of iron. In one instance, it presents a septarian form, and is divided into thin polygonal plates, intersected by a number of cross fissures. The divisions of another are triangular or polygonal, and of a third rectangular, so that in all the beds, their tesselated appearance gives them the artificial aspect of a mosaic brick pavement. Some of the beds of shale also contain nodules of hematite.

About half way between the Salt Pans and the Cock of Arran,* are several beds of red limestone (Section No. 1.) The principal one is of considerable extent, and crosses the beach from the sea in a northwest direction. It is very hard and compact, somewhat granular in its texture, and overlies a bed of red shale, which, gradually hardening as it approaches the limestone, passes imperceptibly into it. This limestone is peculiarly rich in fossils, and large stems of Encrinites; some of them 10 inches long, have been extracted from its surface. The principal fossils of these beds are as follows:—

* The Cock of Arran is a large stone on the beach, forming a well-known landmark to seamen. Formerly, when seen in some positions from the sea, it presented the appearance of a cock in the act of crowing. Some idle or malicious persons have since broken off the head, which now lies on the ground beside the decapitated body.

1. Producta lobata.	6. Spirifer undulatus.
2. do. horrida.	7. do. octoplicatus.
3. do. Martini.	8. Various Encrinites,
4. do. latissima.	Corals, &c. &c.
5. do. Spinosa.	

Part of the surface of the largest bed is very much honeycombed with the perforations of a species of Pholides, the highest of these borings being now about six or eight feet above the tidal level. One of the surfaces of a large square block of limestone, which rests on the underlying shale, is perforated to a great extent, and fine specimens of this phenomenon may be obtained from it: but from the fact of only one of its sides being marked, it is evident that it has been drilled previous to its separation from the parent rock. These perforations must therefore have been made when the rock was submersed in the sea, and it has been subsequently raised to its present position, which elevation probably caused the disappearance of the perforating Testacea. Circumstances of this nature, more convincing than any abstract reasonings, serve to denote to the sceptic in the doctrines of geology, if such there still may be, the great laws which regulate the changes of the surface of the globe. Here we have a bed of limestone composed of shells, which have lived and died deep in the basin of the sea. Their remains, mixed with other extraneous matters, are consolidated into a compact limestone rock, which again becomes the abode of later Testacea, and is now elevated to the surface of the solid earth, again to be decomposed, and washed into the sea, there to share in the composition of other strata now in process of formation.

The preceding group is succeeded by beds of grits, sandstones, white freestone, and shales. Above them lie various beds of red and gray marles, interspersed with ironstone nodules, and alternating with white sandstone; and above this latter group, which may be considered as intermediate between the carboniferous series, and the lower beds of the new red sandstone, lie the undoubted beds of this latter formation. (Section No 1.)

Proceeding along the shore from the Salt Pans, the collective thickness of all the strata of this latter formation, can be seen at one view. It rests conformably above the carboniferous formations, but reaches to a much greater elevation, rising in rounded hills to the height of about nine hundred feet. Its lower beds are principally a coarse conglomerate, alternating with beds of fine sandstone, though nowhere are the embedded pebbles so large as those in the old red sandstone in North Sannox or to the west of Glen Rosa.

These alternations of conglomerate and fine sandstone are common throughout the whole formation, and are not always deposited in regular layers; for if the line of stratification be traced, it will be seen that the same stratum which at one place is a fine sandstone, changes, further on, into a conglomerate. As in the same formation at the head of Glen

Cloy, the conglomerate frequently contains an extroardinary proportion of quartz pebbles, all of which could not possibly have originated in the adjoining formation. In other cases, with an admixture of quartz, the other pebbles seem principally derived from the underlying schists and slates; but here, as in the old red sandstone, there never occurs a single fragment of granite. In general, there is little or no difference of mineral character between this upper formation and the underlying old red sandstone and conglomerate. In colour and structure, and in the nature of the pebbles, they are in every respect similar, but the intervention of the carboniferous series, is of itself sufficient to warrant their distinction into two distinct groups; and the occurrence of cornstone in the higher strata of the lower conglomerate, and of arenaceous grauwacke slate, at its junction with the slate, sufficiently demonstrates its identity with the old red sandstone. It may probably also contain fossils, though they have not yet been discovered. At least, in one instance, a Stigmaria, such as are common in the lower beds of the new red sandstone, has been found in the conglomerate beds on the south side of Brodick Bay, which, it will be remembered, bears the same relation to the carboniferous strata, to the south of the anticlinal axis, as the rocks now under review do to those of the north.

Between the southern limit of the new red sandstone, and the Cock, the rocks overhang the side of the hill in a succession of bold and terraced cliffs.* These cliffs, at a place called the Scriden, have given way and rolled down the hill, encumbering the shore with their massive fragments. The scene here is, if possible, even more magnificent than that at Fallen Rocks, and from the extent of the debacle, it is necessary in following the line of coast, to climb over the ruins by a rough and winding route, rendering the passage, to unaccustomed feet, both fatiguing and difficult. It is an arduous climb to reach the summit of the hill, but when reached, it will be seen that the great mass of the strata has been shifted from its original position, and slid down the mountain, and being rent laterally in many places, as the masses were detached from each other, they are now separated by narrow chasms of great depth. The entire hill has thus been shaken, for, near its summit, where the rocks do not appear to have materially changed their position, a similar rent occurs, not more than a yard broad, and so deep that the bottom is lost in the darkness. The edge is almost covered with heath, and to the unwary passer by, who may be gazing at the peaks of Caistael Abhael, it might form a very awkward pit to fall into. The entire phenomenon thus developed, is in fact a landslip, and the steepness of the incline over which it slid, occasioned the disengagement of the tumultuous ruins which now encumber the beach, and strew the slope of the hill. The immediate causes of this phenomenon, were possibly

* In winter, some of these cliffs on the coast have a peculiarly striking appearance. One of them, the surface of which is constantly wet, by water oozing from the soil above, may be sometimes seen during a severe frost completely covered with ice, like an immense wall of clear glass.

as follows:—Part of the strata which covered the steeper part of the hill immediately above the shore, was probably gradually destroyed by the action of the elements, and by the rains running down its sides, thus leaving an overhanging mass. If springs existed, or in any other manner water insinuated itself between the rocks which composed this cliff, and the underlying strata, the weight of the superincumbent rocks would cause them to slide down the slope of the hill, and the wild and disorderly scene now exhibited, is the necessary consequence of their fall.*

The same strata continue for a little to the north of the Scriden to form the hills on the shore, but at a hollow in the hills, through which runs a stream called, Alt Mhor, the new red sandstone disappears, and in the coast section is followed by schist, which, in continuation of the lower inland hills from Glen Sannox, here extends to the sea, (See Map and Section No. 1.) The schist is here of that description called chlorite schist, and in various places along the shore contains a great quantity of quartz, sometimes disposed granularly throughout the rock, but generally in irregular veins of varying size, not however so thin as to present a laminated structure. Its dip varies from 60° to 64° south, and a reference to section No. 1, will show that this is the prevailing dip of the entire mass of this class of rocks to the south of Loch Ranza. The N.W. dip of the old red sandstone of the carboniferous series and of the new red sandstone to the north of the anticlinal axis, is entirely uncomformable to the dip of the underlying slate and schist, (Section No. 1,) and unless some other reason can be assigned for this unconformability, it is evident that a great and general disturbance, and dislocation of the hypogene strata had occurred, previous to the deposition of the secondary sandstones. This unconformable disposition

* The following account of a similar, but much greater landslip, on the southern face of the Himaleh, is extracted from "Travels in the Himalayan Provinces of Hindostan and the Panjab, by Mr. W. Moorcroft and Mr. G. Trebeck; London, 1841:"—

"About two-thirds up the acclivity of a mountain, about half a mile distant, a little dust was from time to time seen to arise; this presently increased, until an immense cloud spread over and concealed the summit, whilst from underneath it huge blocks of stone were seen rolling and tumbling down the steep. Some of these buried themselves in the ground at the foot of the perpendicular face of the cliff; some slid along the rubbish of previous debris, grinding it to powder, and marking their descent by a line of dust; some bounded along with great velocity and plunged into the river, scattering its waters about in spray. A noise like the pealing of artillery accompanied every considerable fall. In the intervals of a slip, and when the dust was dispersed, the face of the descent was seen broken into ravines, or scored with deep channels, and blackened as if with moisture. About half a mile beyond, and considerably higher than the crumbling mountain, was another whose top was tufted with snow. It was surrounded by others, lower and of a more friable nature. It appeared to me that the melting of the snows on the principal mountain, and the want of a sufficient vent for the water, was the cause of the rapid decay of the mountains which surrounded it; for the water which in the summer lodges in the fissures and clefts of the latter, becomes frozen again in winter, and in its expansion tears to pieces the surrounding and superincumbent rock. Again, melting in the summer it percolates through the loosened soil, and, undermining projecting portions of the rock, precipitates them into the valley. As, however, rubbish accumulates on the face and at the foot of the mountain, a fresh barrier and buttress are formed, and the work of destruction is arrested for a season."

of the schist and new red sandstone, is beautifully exemplified on the beach at Newton Point, where a very small stream called Alt Beithe flows into the sea. The sandstone may be about three hundred yards in length, and only rises a few feet above high water mark. The schist here dips at a high angle to the S. S.W., and on its upturned edges rests the patch of red sandstone and conglomerate dipping to the N. N.E. at an angle of about 40°.

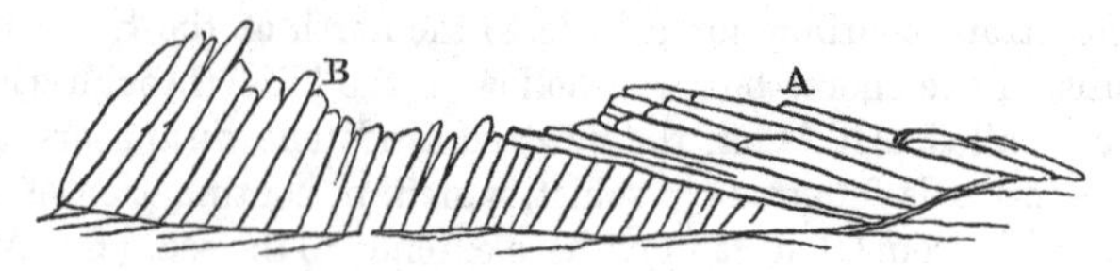

Fig. 12. Unconformable strata of New Red Sandstone at Alt Beithe. A New Red Sandstone. B Chlorite Schist.

Here, therefore, it is evident that the edges of the schist must have been upturned previous to the deposition of the overlying sandstone strata. The same line of schistose cliffs, with the same dip, continues to line the coast which turns into Loch Ranza. It is much contorted and veined with quartz running both parallel to the lines of stratification, and also laterally. These veins vary from an extreme minuteness to a foot or eighteen inches broad. Some masses of quartz, about two feet in diameter, lie on the shore, being evidently pieces derived from the larger veins. In the cliff which, with little interruption, bounds the shore all along from Brodick, there are some sea-worn caves on the coast of Loch Ranza, which, along with those to the south, are further indications of the elevation of this entire line of coast.

There is perhaps no scene in Arran, which so much impresses the beholder with the feeling of solitary beauty as the first glimpse of Loch Ranza. The traveller may perhaps be somewhat fatigued with this protracted journey, as on a still summer evening he rounds the Newton Point. But tired and hungry though he be, and with the very smoke of the little inn curling before his eyes, let him pause for a moment at the entrance of the loch, and seating himself on a granitic boulder, quietly contemplate the placid scene before him. Trees there are few to boast of, and what is pleasanter, there are still fewer strangers, for to the traveller in such a scene, all strangers seem out of place but himself. The sinking sun shines bright on the gleaming peaks of Caistael Abhael and Ceum Na Cailleach, where the shadows of the ragged scars and deep hollows of the winter torrents, mingling with the lights brightly reflected from the projecting rocks, form a hazy radiance, which more obscures than illuminates, the shady recesses of the rugged Corries. The tide is at its full, and the lazy sails of many a lagging fishing boat, the image of the ruined tower and of the green hills around, lie calmly reflected in the unruffled waters:—

" The lake return'd in chasten'd gleam
The purple cloud, the golden beam;

> Reflected in the crystal pool,
> Headland and bank lay fair and cool,
> The weather-tinted rock and tower,
> Each drooping tree, each fairy flower;
> So true, so soft, the mirror gave,
> As if there lay beneath the wave,
> Secure from trouble, toil, and care,
> A world than earthly world more fair."

But it is in a cold February evening that the pleasant solitude of the place will be most esteemed. There, seated at a blazing peat-fire, as the geologist extends his notes, or arranges his specimens, after his day's work, he will hear the piercing wind whistling down Glen Chalmadael, and the narrow pass of Glen Eisnabearradh, then dying away as it reaches the wider expanse of the loch, to be again renewed by a louder and a shriller blast. And as he loiters to the door to speculate on the probabilities of the morrow's weather, he may chance to see the burning heath, like the beacons of old, blazing on the hills around, and faintly gleaming on the far distant headlands of Argyleshire.

Having described the geology of the coast between Glen Sannox and Loch Ranza, it is now necessary to notice what appearances are worthy of observation in the inland section. It is at once evident that the alluvial plain of Loch Ranza, has been gained from the sea by the encroachments of debris washed down from the hills by streams which now wind through the plain where the sea once flowed. The long expanse of shingle which is covered by the flood-tide, may, in the course of ages, be dry and fertile land like the surrounding fields. The peninsula, on which the castle now stands, is probably a bank formed by the debouchement of some ancient water into the sea, and on the west coast, new banks may be seen in process of formation and gradually encroaching on the bed of the loch. However distant, the period may yet come when the entire basin will be silted up, and the line of coast run between the points at the entrance of the loch. It is only in bays or lochs where such occurrences take place, for there the deposits of the streams may rest comparatively undisturbed; whereas, where a river flows into the sea on a nearly straight line of coast, such as at North and South Sannox, the alluvial deposits must neeessarily be small, owing to the transported matter being washed away by the waves as soon as it is deposited.

The elegantly conical hill of Toirnaneidnoin,* forms the southern boundary of Glen Ranza, and on the west side of that hill may be seen the celebrated junction of granite and schist which bears the name of the place where it occurs. Two large veins proceeding from the mass of granite penetrate the schist in a sloping direction, towards the top of the hill. In the main junction of the granite and schist, the granite is pretty large in the grain. The cooling and crystallizing process was therefore

* Or the Mount of Birds' Nests.

gradual; but this crystalline granite is generally separated from the schist, by a narrow compact strip indicative of more rapid refrigeration. In the two veins which penetrate the schist, this phenomenon is particularly observable, the granite being small-grained, and of a harsh sandy texture. The parallelism of the layers of stratification is much disturbed, and these contortions generally form curves of considerable magnitude, though there are many on an extremely minute scale, interlaminated with layers of quartz, and forming angles so acute as 33°. At this junction, as at most of the others, the schist is much interpenetrated by quartz veins, both parallel with, and at various angles to the plane of stratification. The smaller veins are, however, generally perfectly conformable, alternating with the strata, in regular laminæ, and yielding to every bend and contortion in the stratification. It is unnecessary at present to enter on the theory of these veins, but the remark may be repeated, that generally they increase in number where the slate joins the granite.

As both sides of Glen Eisnabearradh are of granite, and those of Glen Chalmadael of slate, it is evident that the line of junction must stretch along the west side of the latter Glen. It is, however, totally obscured by an extensive moss.

The rocks which encircle Glen Chalmadael are composed of every variety of the Arran hypogene strata; argillaceous chlorite and micaceous schist apparently intermingle in inextricable confusion. On the hills above Cock Farm are quarries, now deserted, where formerly both the chloritic and argillaceous varieties were wrought. Farther south the schist assumes a granular texture, containing numerous small rounded nodules of quartz, which sometimes present the appearance of an imperfect crystallization. A little felspar is occasionally sparingly intermingled, and an inexperienced observer might at first, in some instances, mistake it for a variety of granite.

After reaching the end of Glen Chalmadael the descent towards the south is by a little hollow called Glen Dubh, through which flows a small stream tributary to North Sannox water. Several rivulets descend into Glen Dubh from the west, and in the most northerly of these the junction of the granite and slate may be seen, exhibiting the usual phenomena. Farther south a junction, which seems never before to have been described, takes place in the bed of the main stream of Glen Dubh. It is exceedingly interesting, illustrating in itself almost all those peculiarities which are generally to be seen only by the inspection of numerous examples. The contact of the main bodies of granite and slate is first visible in the channel, about half way between North Sannox and Glen Chalmadael. It crosses the bed of the burn in a north and south direction, after which it is lost in the moss. But the boundary line must skirt along close on the west bank; for, about a hundred yards below the main junction, a vein of granite, five feet broad, crosses the stream, and again, thirty yards lower down, a second vein, one foot broad, traverses the slate in the same manner.

What most forcibly arrests the attention at this junction is, the extraordinary alteration to which the slate has been subjected by its proximity to and contact with the granite. Not only is the plane of stratification contorted and veined with quartz, but between the granite and the small vein, which is about a hundred and thirty yards below the main junction, the entire mass of slate presents a whitish and, if the expression be allowable, a semi-granitic appearance, distinctly showing the partial or entire fusion of the slate, by which it has become intermixed, or partially blended with the melted granite. Below the lowest granite vein, the slate, as it recedes from the granite, gradually assumes its natural colour and appearance. This view of the subject is borne out by the circumstance that at the main junction* in this stream, the granite is peculiarly coarse and crystalline in its texture, from which it may be inferred that it cooled there very slowly, a result arising from the partial or complete fusion of the neighbouring slate. The five feet vein is, for the reasons already stated, of a fine and compact texture, but without undergoing any particular alteration in its colour or other characteristics. The granite of the one foot vein has undergone a greater change, being of a blueish colour—finer and more compact than the other vein, and extremely brittle. This alteration in its mineral character is probably caused by its more intimate association with the penetrated slate. As a general rule, it has been observed that wherever the granite comes in contact with the slate, it becomes more brittle and easily fractured, a phenomenon also resulting from the increased rapidity of refrigeration.

About two hundred yards above the confluence of Glen Dubh and North Sannox waters, another junction may be seen in the hollow of North Glen Sannox; but, as the observer must now be somewhat familiarised with these appearances, it is unnecessary to describe it, though to the beginner repeated observations will always be instructive. In the first tributary of North Sannox water, to the south-east of its confluence with that of Glen Dubh, the junction is again apparent at a considerable height on the hill; and here, as at those formerly noticed, the slate dips 74° to the south, so that, instead of resting on the granite, its west edge abuts against it.

The descent of the slate† into Glen Sannox has already been described in accounting for the disappearance of the slate in that Glen.

The scenery of North Glen Sannox is of a simple and unobtrusive character. Here the loneliness is as complete as when, in the recesses of the southern glen, the mind of the spectator resigns itself to the mysterious feelings of dread, which associate themselves with the stern grandeur of the surrounding objects. There the idea of desolateness originates in the absolute isolation of the beholder from all the sights

* By main junction is always meant the contact of the principal masses of granite and slate, or schist, not that of the granitic veins.

† In a slight hollow at the foot of Suithi Fheargus, between the last named junction and Glen Sannox, there is an echo which repeats a shrill cry with great distinctness six several times.

and sounds which bespeak the existence of man; while the ruined cottages in the northern glen tell a different tale, showing that its green solitudes once afforded a home to many families.

North Glen Sannox was once what, in a highland district, is considered populous. Much to their own advantage, its inhabitants were forced to emigrate to America. Their huts were of the poorest description, and, as fishermen, they earned but a scanty and uncertain subsistence.

Twenty or thirty years ago, the geologist, as he came down the glen, armed with the implements of his calling, might espy the lazy fishers, on a fine summer afternoon, lounging at their doors, to enjoy the luxury of an evening pipe; whilst the women and children salute, with a wondering stare, the insane stranger, who puts himself to so much trouble and fatigue to collect a bag of useless stones; nor do they ever attempt, by word or action, to call off the pack of yelping curs, which, snarling at the traveller's heels, pursue his retiring steps; their loud and prolonged barking, as he leaves them in the distance, marking their anxiety to expel the suspicious intruder. But the scene is different now. Green spots, clothed with a close cropped herbage, and still bearing witness to the marks of the plough, surround each ruined clachan. The hazel and the fragrant birch, the ash and the charmed rowan, fringe the banks of the stream, or mark the remains of the little garden enclosures; and mingled with these may be seen the white blossoms of the gnarled elder, famed of old for its irresistible power in scaring the midnight witches from the neighbourhood of lonely dwellings, and counteracting the malicious pranks of the fairies, who, it is well known, still inhabit these desert wastes!

Granite Hills from Lagan Hill.

Conical Boulders.

CHAPTER IV.

TRAP DYKES FROM CLACHLAND POINT TO LOCH RANZA.

HITHERTO little allusion has been made to the numerous trap dykes, which intersect the various formations of the district which has now been described. To have mentioned each as it occurred, would have occasioned a constant interruption of the consideration of the formations under review, to prevent which, it seemed better to delay any notice except of the more striking dykes in this particular district, till a certain line of coast had been examined.* The geologist may, therefore, occasionally glance at this chapter during his progress over the ground already described.

Besides the great pitchstone dyke already noticed, there are at least eight greenstone dykes between Clachland Point and the turn of the Corriegills coast into Brodick Bay. The first of these is immediately north of Clachland Point. It is six feet broad, points to the N. E., and from its decomposing more rapidly, is slightly below the average level of the new red sandstone. Near the north end of the great pitchstone dyke, is another dyke running E. and W., of great size, and, unlike the last, it

* In the year 1839, M. L. A. Necker made an admirable survey of the trap dykes in Arran, more particularly of that district lying between King's Cross Point and Loch Ranza. The results of his very laborious observations, he laid before the Royal Society of Edinburgh, on the 20th April, 1840, now published in a pamphlet entitled, "Documents sur les Dykes de Trap d'une partie de L'Isle de Arran." In a number of instances I personally confirmed the accuracy of his statements, which are generally so full as to leave no room for further remarks; and I am therefore indebted to his valuable paper for many of the remarks in this work on that subject.

projects above the surface, owing to its superior resistance to the effects of the elements. Close beside, and on the west extremity of this dyke, is one of great length, and six feet broad, running into the sea in a due north direction, being partly above and partly sunk below the surface. Near its north extremity, and close on the coast, another sunk dyke, one foot broad, diverges to the north-west. Immediately to the north of these, is a mass of claystone, accompanied, on the north border, by a wedge-shaped mass of pitchstone, covered with spherulites. On the bank immediately to the south-east of the first house to the east of Low Corriegills, is another dyke, one foot broad, pointing to the N. W., which is succeeded by two others, the south one of which points to the N. E., and that at the turn of the bay due east. It is of a very dark greenstone. Somewhat west from the point, are two sunk basalt dykes, (N. N. E.,) one of which is twelve feet broad. A little further west, and immediately to the east of a second mass of claystone, is a greenstone dyke, two feet broad, (N. E.) On the bank the high part of this dyke penetrates a conglomerate, but below it is bounded on one side by claystone. On the high bank it is raised above the surface like a wall, but on the coast it is sunk between two elevated bands of sandstone, each about a foot in thickness. Between this point and the east side of the Springbank bridge,* are seventeen dykes, varying in breadth from one to thirty feet. The six central dykes form the largest of this range, varying in breadth from twenty to thirty feet. The fifteen most easterly dykes point more or less to the north-east, and the remaining two to the north-west. They penetrate the new red sandstone or conglomerate, and are all composed of greenstone, with one exception, the sixth from the eastern dyke being a felspathic trap. The third from the east is elevated above the surface, the rest being all more or less sunk. In some of them the sandstone is much hardened and whitened in the neighbourhood of its junction with the trap, thus evincing the alteration to which the sandstone has been subjected, by contact with a heated substance.

Another dyke, immediately to the south of Springbank bridge, penetrates the conglomerate from east to west. Somewhat to the south-west of this dyke, where the Corriegills road diverges from that which leads to Lamlash, there is a range of seven dykes, formed of what may perhaps be called syenitic greenstone. They vary in thickness from one to eight feet, and run more or less in a north-east direction. With the exception of the east dyke, which penetrates a hard sandstone, the remaining six, taken together, seem to form one great dyke, composed of alternate bands of greenstone and claystone porphyry; the new red sandstone conglomerate and red marles which it traverses, being only visible at the east and west extremity: it forms the bed of a little rivulet running parallel to the road. A little beyond these dykes is another, which crosses the road; and another, very large, and composed of green-

* This bridge is a little west from where the Brodick road turns up the hill towards Lamlash.

stone porphyry, also crosses the road at the most westerly house of high Corriegills. On the Lamlash road, four dykes may be seen penetrating the new red sandstone: some of these are much decomposed, but as they exhibit nothing of peculiar interest, it is unnecessary to describe them particularly. Going westward, a little beyond Springbank bridge, a small sunk greenstone dyke, one foot broad, penetrates the conglomerate, (N. 35° E.,) which is succeeded by another, three feet thick, composed of an amygdaloidal felspathic trap, (N. 10° W.,) also sunk, the penetrated conglomerate being slightly hardened at the junction. To the west of this is another, six feet broad, (N. 20° E.,) followed by a dyke of the same size. This dyke is worthy of observation, as it divides into two branches, the principal of which diverges to the south-east. The next is the largest on this range of coast, being thirty feet broad, penetrating a whitened conglomerate, (N. 20° W.,) which is much hardened at the junction.

It will now be necessary to diverge to the left into Glen Cloy, where there are a few dykes, which, however, do not present any more remarkable appearances than those already described. None are visible for a considerable distance up the bed of the stream. In the bed of a small tributary which flows into Glen Cloy Water, the largest dyke yet noticed penetrates the new red sandstone conglomerate in a northerly direction. It is fifty feet in breadth, and is decomposing into balls, like the trap of the remarkable dyke which has already been described as penetrating the granite of Ciormhor. Immediately above the confluence of the two waters at the mouth of Glen Dubh, is another dyke, (N. 10° W.,) which, it may be observed, makes no alteration on the sandstone through which it passes. On the north side of Glen Dubh, a sunk greenstone dyke (N.) fills the hollow of the deep ravine which separates Sgian Bhain from Tornadearc. This forms a fine example of the more rapid decomposition of the igneous than the stratified formation which it penetrates, originating a hollow, the bottom of which is here filled by the dyke. Its breadth varies from ten to twelve feet. Immediately to the north-west, there is another about forty-five feet broad (N. W.) slightly sunk below the surface. At the south-west base of Tornadearc, a compact greenstone dyke, six feet wide, (N. 80° W.,) penetrates, without altering either the colour or the texture of the new red sandstone conglomerate. Another may be seen on the plateau which bounds Glen Cloy on the south-west.

It was already stated, that a trap dyke penetrates the porphyry on the south side of the summit of the Windmill Hill. It is extremely compact, and difficult to fracture, and appears to have slightly darkened the colour of the penetrated rock.* Two small dykes penetrate the red micaceous

* I may here state, that it is impossible for any person always to find all these dykes by description, or indeed any limited formation which may be sought for. This dyke I observed the first time I visited the Windmill Hill, and near it I found a particularly beautiful porphyry. Since then, I have twice explored the summit, but have never since been able to discover either the dyke or the porphyry.

sandstone of the carboniferous series in Glen Shirrag, one (N. 60° E.) near the north-east base of the Windmill Hill, at the top of the rivulet which flows into Glen Shirrag, and the other (N. 70° E.) immediately below Brodick church. They are both parallel to the stratification of the penetrated sandstones, and though of different thickness, and dipping at different angles, it is probable that they are a prolongation of the same dyke. In Glen Rosa, a little beyond the farm, are two small dykes, much decomposed, running nearly north and south, and penetrating a micaceous rock of the old red sandstone formation. A little below the confluence of Glen Rosa Water with Garbh Alt, three trap dykes penetrate the talcose schist, which here forms the bed of the stream. As the schist is here near its junction with the granite, the alteration it has thus undergone, renders it impossible to say whether or not it has been at all changed by contact with the trap. Any notices regarding the ordinary trap dykes in the granite, are reserved for consideration in connection with other phenomena relative to that formation in Chapter VI.

The large dyke in Cnocan Burn has already been described in the account of the ascent of Goatfell. In the sandstone on the north side of Brodick Bay is a greenstone dyke, which, forming a right angle, pierces the sandstone both parallel with and at right angles to the plane of stratification. Its first direction is N. 22° W., and the second N. 45° E., and its angle of inclination about 80° S. 68° W., and again 45° S. W. The sandstone is somewhat hardened and altered by the contact. Immediately to the north is another dyke one and a half feet broad, (N. 70° E.) which does not alter the penetrated bed. A little further north is another one foot broad, which is succeeded by a large curved dyke fifteen feet broad, its direction being first N. 10° and then 20° W. The inclination of the beds traversed is variable, being on the east side of the dyke partly N. 20° W., while the strata to the north and south dip in a north-east direction. At one end the sandstone is very white on both sides for about twenty-five feet from the junction, but elsewhere it is only close on the junction that its colour is altered. Further north are two greenstone dykes from four to six feet broad, one of which is slightly curved. About a hundred paces to the south of the gate at the north end of the castle wood, is a long dyke three feet broad, and which, like those already noticed on this coast, is also sunk. Its direction is N. 25° E., thus precisely coinciding with the large dyke twelve feet broad, the third from Corriegills Point, of which it may probably be a prolongation.

From this point to Little Port, (or Port Na Claoch,) no dykes are visible on the coast, but at the summit of Maoldon a large dyke penetrates the carboniferous sandstone in a due north direction, dipping to the west at an angle of 85°, the general inclination of the sandstone being to the south-east. At Screeb Quarry, to the north of Maoldon, two greenstone dykes penetrate the lime and sandstone, (N.) the one being ten, and the other four feet broad. In the bed of the stream, to the north of Maoldon, are two narrow dykes exhibiting appearances of much interest. The most easterly is composed of a black porphyritic basalt, and vertically pierces the carboniferous sandstone, ending in a bed of basalt of the same texture, which has overflowed the underlying bed of red shales and sandstone to a considerable depth. The red colour of the sandstone is not altered at the point of contact, but at some places it is somewhat indurated, while elsewhere its texture remains the same. A little further up the bed of the stream, a compact trap dyke, two and a half feet broad, pierces without altering the same set of strata; but it is also continued through the overlying basalt, from which we may conclude that it is of later origin. It is a little below where the burn forms a beautiful waterfall in a cave, which has been worn by the stream in the sandstone and conglomerate beds which it traverses. The basalt is rapidly decomposing. It contains a great proportion of hornblende, and encloses numerous veins of carbonate of lime. It is also partly amygdaloidal, and, indeed, in many respects, bears a strong resemblance to the large dyke already mentioned as associated with the limestone strata at Corrie. There are four dykes on the coast between the White Water and the village of Corrie. The second north from the mouth of the stream is very small, the others vary from one to four feet in breadth. The last, which is immediately to the south of the village, does not in the least degree alter either the colour or structure of the sandstone. There are three dykes in the Lime Quarry, which, along with the great bed associated with the strata intervening between the quarry and the old red sandstone, have already been noticed in the description of these formations.

No dykes have yet been observed between Corrie and the old march of the farms of Sannox and Lagantuin. Between the great ecroulement at the Fallen Rocks and this point, there are seven dykes. Two of them are a little to the north of the boundary line. They are of greenstone, and are both of considerable magnitude, and, being parallel, penetrate the old red sandstone from east to west. Two parallel basalt dykes further north take a direction N. 45° W. The most southerly is forty-five feet, the other eight feet broad. These are again succeeded by other two of smaller size, also parallel, and penetrating the sandstone from east to west. All of these dykes are more or less sunk and nearly vertical. A little south from the Fallen Rocks, a vertical trap dyke, several paces broad, may be seen penetrating the old red sandstone conglomerate, and its acompanying white and violet coloured argillaceous beds. This, it will be remembered, is near the borders of the old red sandstone and carboniferous series. To the north of the Fallen Rocks, are the masses of traps parallel to the plane of stratification of the conglomerate. These have been already described in the account of the northern carboniferous series. Between this point and the Salt Pans is a narrow sunk greenstone dyke one and a half feet broad, direction N. 10° W., dipping 70° N. 10° W., between the black calcareous fossiliferous beds and the red limestone, with its accompanying green shales. The inclination of the stratified deposits is here 50° N. 20° E., so that the dyke penetrates them nearly at right angles to the plane of stratification. Immediately south from the Salt Pans, very feeble traces of a sunk porphyritic greenstone bed may be seen traversing the strata from east to west, and which appears to be parallel to the penetrated beds, lying below a whitish grey sandstone, and above the black shale already noticed as being so rich in impressions of plants. To the north of the coal a similar dyke penetrates the strata, and as these two dykes dip in opposite directions, and intersect each other, (See Section No. 1,) they thus cut off a small triangular portion of the coalfield, the only part of it in which the works were prosecuted.

Immediately to the south of a little artificial harbour, formerly used for the shipment of blocks from a new red sandstone quarry, near where it overlies the carboniferous series, is a long black basalt dyke, which is slightly curved at various places, and may be traced (N. 10° W.) to the port, along a space extending about 650 feet. It is about four feet broad. Beyond this is another on the steep bank of the hill. At the Cock of Arran, a small greenstone dyke, one foot broad, stretches N. 10° E. Towards the south it branches into two divisions, and is in several places bounded by a band of ferruginous clay of a bright red colour. Close to the north, a dyke of the same size, and having the same direction, divides into three ramifications, the western branch being somewhat curved to the north. A little to the north of the Cock, a dyke composed of a fine greenstone porphyry, containing well developed crystals of felspar, is elevated above the surface like a wall. It is seven feet broad, and stretches across the shore N. 45° E. At the bottom of the north side is a very small dyke slightly elevated above the surface, and perfectly parallel to the last. It is composed of a light coloured trap, containing a considerable proportion of quartz.

There are no trap dykes beyond this, in that district where the schist forms the coast, till we arrive at that little patch of unconformable new red sandstone at Newton Point. Here, and in the neighbourhood, there are five sunk dykes penetrating the schist and the sandstone. The most easterly is fifteen feet broad, the others vary in breadth from two to eight feet. The first four are parallel, (N. 20° W.,) but the west dyke stretches first N. 25° W., and then N. 50° E. The dip of the schist is 68° S. 20° E., and beyond the curve the dyke sinks beneath it. On the east side of Loch Ranza, opposite the ruined castle, a greenstone dyke, four feet broad, penetrates the beds of contorted slates, and talcose schist, pointing N. 10° E.

It has been already remarked, that when a penetrating trap dyke is more durable than the penetrating strata, the rock which it traverses being worn away by the action of the elements, the trap dyke is left above its surface in the form of a wall. But it sometimes also happens, that, being more perishable than the rock which it pierces, the trap decomposes and leaves a hollow in the cliffs, bounded on each side by a perpendicular wall of sandstone. Among many, two well marked examples of this occur in the cliffs of the Corriegills shore; the accompanying figures illustrate these phenomena.

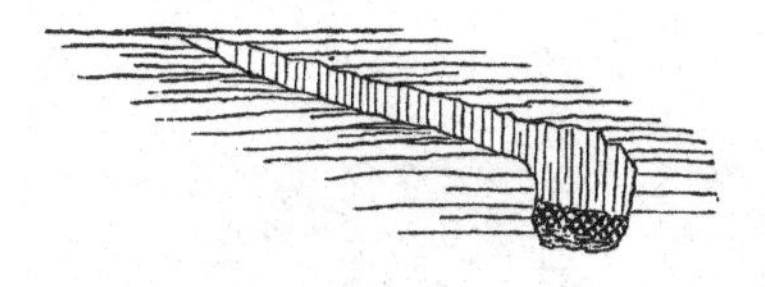

Fig. 15. Sunk Trap Dyke.

Fig. 16. Raised Trap Dyke.

Corrie an Lachan.

CHAPTER V.

LOCH RANZA TO CATACOL—QUARTZ VEINS IN THE SLATE AND JUNCTIONS IN CATACOL BAY—LINE OF COAST TO GLEN IORSA—OLD RED SANDSTONE—SINGULAR VEINS NEAR TORMOR—KING'S HILL AND CAVES—DRUMMEDOON POINT—TRAP DYKES—HILLS TO THE SOUTH OF SHISKIN ROAD—CLACHAN GLEN CARBONIFEROUS ROCKS—LEAC A BREAC—SLAODRIDH WATER—PORPHYRIES AND INDURATED SANDSTONE—TORLIN WATER—FROM LAG TO BENAN HEAD—STRUYE ROCKS—NEW RED SANDSTONE—DISTRICT BETWEEN BENAN HEAD AND THE DIPPIN ROCKS—EIS A CHRANAIG—FROM WHITING BAY TO LAMLASH—GLEN ALASTER—LIMESTONE THERE—REASONS FOR REFERRING THE STRATIFIED FORMATION OF THE SOUTHERN DISTRICT TO THE CARBONIFEROUS AND THE NEW RED SANDSTONE SERIES—HOLY ISLE—PLADDA.

The most interesting and varied portion of Arran has now been described, embracing the principal phenomena of all its formations; including the underlying granite, the overlying igneous rocks, and the stratified series, from the schist and slates to the new red sandstone, which is the latest stratified formation in Arran.

A reference to the map will show, that the greater part of the northern half of the west coast is entirely composed of schist, resting directly on the central granitic nucleus. With the appearances which characterise the contact and approximation of these rocks, the observer is already familiar. It is therefore unnecessary to enter on a minute account of this district, though, as it is always of great importance to familiarise the eye with the various appearances which it presents, it may still be traversed and examined by the student with great advantage.

The district between Loch Ranza and Catacol is partly composed of micaceous, partly of argillaceous schist. If any boundary exist, it is impossible to define it, as the rocks frequently intermingle and alternate with each other, and are deeply covered with soil. The general dip of the strata is towards the granite, about 45° S. It is all more or less contorted, and much veined with quartz. In the bed of Uisge Solus, or the bright water, which descends from the east into Catacol Bay, a well defined instance of this may be seen. The strata are of argillaceous schist, and are intersected by numerous quartz veins, varying from the eighth of an inch to a foot in width, and generally parallel to the waving lines of stratification. Immediately to the south-east of this burn, in the side of a hill called Maithaic Uan, the granite may be seen in veins penetrating the vertical strata of clayslate in a similar manner to the junction at Toirnaneidnoin. This bay also presents some interesting phenomena with regard to the formation of alluvial soils, and the silting up of districts formerly occupied by the sea. It is at once evident that the streams which now flow through the alluvial plain of Catacol, are working their way through matter which at a former period they deposited; but the bay is now so nearly silted up, that the debris cannot be expected to encroach much farther on the sea. Catacol water formerly entered the sea considerably farther north than at present. In a season of heavy rains it burst through the middle of the bank, thus destroying a small but convenient harbour for fishing boats. The old channel may yet be traced.

On the summit of Maŏl Na Leaca Sleanhain, which bounds Catacol to the south, is a rounded eminence, characterised by a remarkable junction of the granite and schist. The granite penetrates the latter rock in a number of irregular narrow veins, both of these formations being again penetrated by small veins of a dark coloured syenite. Here, therefore, are three distinct periods: the deposition of the schist,—its subsequent penetration by the granite, and last, the intrusion of the syenite into both.

The whole of the schistose strata, from Catacol to South Thundergay, is perhaps more contorted and veined with quartz than in any other part of Arran. Two remarkable masses of stone, which have fallen from the cliff, stand on the shore at North Thundergay, (see page 73.) From these the place derives its name, and they exhibit in a remarkable degree the phenomena noticed above.

But before descending to the coast, let the geologist turn aside to see a solitary mountain tarn, in the silent recesses of Beinn Mhorroinn. This little sheet of water is by far the most picturesque of all the lochs of Arran, and is situated deep in a hollow, called Corrie an Lachan. The place is perfectly lonely; not a tree is near; and except the brown heath on its margin, and a few stunted rushes by the brook, the surrounding hills are almost bare of vegetation. The water is dark and deep, and the stormy blasts of the mountain never reach its still and unruffled surface. From its edge, on all sides but that towards the sea, rise the naked hills, whose sides are either formed of massive granite blocks, which, though surely yielding to decay, yet offer a stronger resistance to the destroying influences of time than the softer portions of the mountain, where the decomposing rock may almost be seen slowly crumbling away.

A remarkable feature of the granitic hills of Arran, is the Corries, (see page 48.) These may be frequently observed in the ridge between Brodick and Sannox, and in the hills of the interior. They generally present the appearance of a volcanic crater, part of one side of which has disappeared; and the masses of granite which compose the encircling hills, are frequently arranged in layers diverging from the centre of the corrie according to the angle of inclination of the hill. For obvious reasons, it will be evident to the most inexperienced observer, that there is no analogy between the corries* and modern volcanic craters; and it is probable that they owe their origin to the softer nature and earlier decay of the rock, with which at remote periods, they may even have been nearly filled.†

The entire line of coast from Loch Ranza to Iorsa Water is bounded by an elevated cliff, the foot of which is raised above the tidal level to about the same height as that on the east coast. The greater part of the strip of cultivated land between this cliff and the sea is evidently an ancient beach, being still almost as stony and bare of soil as the modern shore. The schist is here and there intersected by dykes, which it is unnecessary to particularise. It may also occasionally be found sparingly intermingled with common slate, as at Mid Thundergay, and near the source of a small stream which flows southwards from the base of Beinn Mhorroin. The approximate conterminous boundaries of the granite and schist, may be traced round the base of this ridge, by marks formerly indicated, and from thence through Glen Scaftigill and Glen Iorsa, and across the flat table land through Loch Ghnuis, and from thence round the base of Beinn Ghnuis to Glen Rosa, where the schist changes into the ordinary clayslate.

To the south of Glen Iorsa, the strata rise on the coast in low cliffs, forming the west extremity of the great old red sandstone band, which here crosses the island to Brodick Bay. It is throughout composed of the usual alternations of sandstone and conglomerate. On the north side of the Shiskin road, near the top of Glen Laodh, the old red

* Corrie or cauldron?

† May not even the great glens owe their origin to the same cause.

sandstone may be seen to have been overflowed by a blueish syenitic rock, which is a continuation of the great masses to the south. Farther west, this sandstone is lost under the large alluvial plain, through which flow Mauchrie and the Black Waters. This alluvial tract is bounded on the west by the sloping eminence of King's Hill, which is partially composed of the common white sandstone. From its colour, and the circumstances of its succeeding the old red sandstone, and being in the same line with the carboniferous formations further west, and in default of other evidence, it is here referred to that formation.

On the coast of Mauchrie Bay, near Tormor, there is a remarkable pitchstone vein, which has been so admirably described by Professor Jameson, that no room is now left for any additional remarks. The following minute notice is therefore here transcribed, from his Mineralogical Tour in the Scottish Isles:—

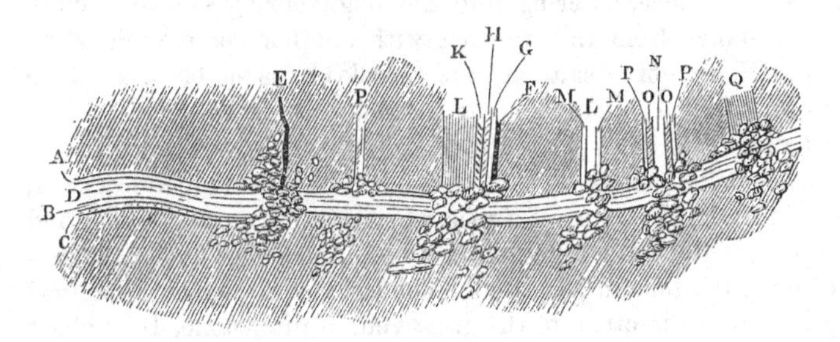

Fig. 18. Pitchstone vein, Tormor.

" The great vein of green-coloured pitchstone, D, as it rises from the sea has a considerable inclination to the horizon, is slightly bent in its course, and traverses the common red coloured argillaceous sandstone.

" Upon the side of the vein next the sea, there is a layer, A, of a substance which appears inclined at an angle of 60°, dips in the same direction with the pitchstone D, and has a similar curve. It is not unlike a compact sandstone, but is probably of the same nature with B, on the opposite side of the vein, only more altered by the action of the weather and the sea. Upon the opposite side of the pitchstone, we observe a layer, B, which appears to be of the nature of hornstone, or rather verging to quartz. Immediately beside it there is a thin layer of basalt, C, which is decomposing in balls; and this again is bounded by the common sandstone strata. The vein continues thus for about twenty yards, when the layers, A,B,C, appear to come nearly horizontal, and soon they disappear under the debris. Further on, where the pitchstone is almost free from the covering of debris, it appears to be bounded on both sides by the common argillaceous sandstone; yet this is doubtful, as there may be small portions of the stratulæ, which the debris prevents us from observing.

" At a little distance from where the sandstone appears to form the side of the great vein D, we observe E, which is a vein of rock similar to that of B, is from six to eight inches wide, and is waved in its course. At some distance from this, there is a vein of basalt, P, about five feet wide, running nearly east and west. The next vein we meet with is about thirty feet wide, runs N.W. and N., and N.E. and E., which is nearly in an opposite direction to the great vein. Upon one side there is a layer, F, of a wax-coloured substance intermediate between hornstone and pitchstone; next is a layer, G, of high olive-green coloured pitchstone about two feet wide; again we have a layer, H, of about half a foot wide, of the same pitchstone hornstone as F, then a layer of indurated clay, K; and after this the whole vein is formed of basalt, L. The sandstone which bounds this vein, in place of being red, the usual colour, is partly a yellowish-white colour. I endeavoured to discover its junction with the great vein, D, but without success, owing to the great covering of debris. I observed it, however, on the opposite side of D, but at a distance, entering into the neighbouring sandstone cliffs. At a little distance from this, we met with another remarkable vein: the sides M M, are of basalt, but the middle L, is of breccia. Still nearer to Mauchrie Bay another curious vein is to be seen: it is about eight feet wide; the sides P P, are of fine white-coloured argillaceous sandstone; next the two layers, O O, of basalt, which decomposes in balls; and the middle N, is formed of a rock which has crystals of felspar and rounded pieces of quartz, immersed in a base that seems one of the gradations from pitchstone to hornstone. The last vein, Q, which I observed running in a cross direction, to the great vein of pitchstone, D, is about ten feet wide, and entirely composed of green-coloured pitchstone."

The cliff to the south of this vein is principally composed of sandstone, in which are a number of water-worn caves, one of which, the well known "King's Cove," is famed throughout the island as having once been the residence of the patriot Bruce, and, in earlier times, of the renowned Celtic hero Fingal, whose image, with his mighty hunters and their dogs, may still be seen rudely sculptured on the walls.*

Many masses of claystone porphyry are associated with the sandstone of this cliff, and both of these rocks are interpenetrated by trap-dykes. One of these has forced a passage between the sandstone and porphyry for some distance; after which it branches into two divisions, one of which pierces the porphyry, and the other continues to divide it from the sandstone. Other veins traverse the porphyry both perpendicularly and also somewhat horizontally. On the cliffs around King's Cove, there are several masses of green-coloured pitchstone, but

* I trust that any one who visits this cave in the hope of seeing these relics, will have eyes and imaginations clear enough to decipher their meaning. Mr. Headrick enters into the controversy with great spirit, and by a skilful analysis of the traditions respecting Fingal or Fioun, and several laborious arithmetical calculations, clearly proves that he was at least eighty feet high. To set the question for ever at rest respecting the fact of his occasional residence in Arran, I was credibly informed by my guide, that in his memory a pot containing a number of antique coins and a silver chain was dug up by an old woman in Catacol, and this chain, it is asserted, was the very chain, with which Fingal bound up for the evening his dog Bran.

from the nature of the ground it is difficult to determine whether they are dykes or overlying masses. Several trap-dykes penetrate the sandstone cliff near the caves, and the basaltic promontory of Drummedoon. The rocks which compose this picturesque promontory are partly composed of claystone porphyry, containing crystals of felspar and small granular crystals of quartz, imbedded in a base of a light grey colour. This base frequently varies to a dark leaden blue shade, containing crystals of what is sometimes called glassy felspar. In some places it assumes the construction of what may be best designated as a species of syenite, being composed of quartz and felspar but without the hornblende. These formations assume a rudely columnar form, and rest on a thin stratum of white sandstone, which intervenes between them and the principal strata of underlying red coloured sandstone. Three dykes cross the beach in an east and west direction, a little to the south of Drummedoon. Two of these consist of a beautiful dark coloured porphyry, containing large crystals of glassy felspar, which in many instances have disappeared, the vacancies being replaced by a cream-coloured carbonate of lime. In some of the crystals of felspar, this displacement seems as if now in process. Has this singular phenomenon yet been accounted for?

The lower part of the hills to the south of the Shiskin, and on part of the south side of the Brodick road, are principally composed of sandstones and conglomerates, but further to the north-west, approaching the String, the southern hills are formed of various syenites, and rocks sometimes of an intermediate nature between that and the other igneous formations. It is impossible always to give these varieties specific denominations.

In ascending Clachan Glen, the strata partly assume a conglomerate form, alternating with beds of sandstone, but further up the glen we meet with disjointed masses of a pure compact limestone containing pebbles of quartz, thus forming a calcareous conglomerate, analogous to that where the carboniferous formations join the old red sandstone, at the Fallen Rocks and near Corrie; on which account, and from its proximity to another bed of fossiliferous limestone, afterwards to be noticed, the strata with which it is associated are now referred to the carboniferous series. The upper beds are frequently of a purer lime, which, however, is destitute of fossils. It also contains a small proportion of sulphate of barytes, and on being fractured, the surface sometimes presents the very common phenomenon caused by the *efflorescence* of oxide of manganese. The entire hollow of this glen is composed of various strata of sandstone, and the steep banks are crowned with igneous formations, to the varieties of which it is impossible always to assign a locality from the want of proper points of reference. They are also so intermingled, and blend into each other by such imperceptible gradations, that, as already stated, it is in many instances impossible to assign them any distinctive name. Except in the case of dykes which intersect the sandstone strata on the shore and elsewhere, the southern igneous formations may, as a whole, be classed as overlying irrupted rocks, resting on the stratified formations, which, in the southern district, are therefore the lowest geological formations.

On the coast, to the south of Black Water, the sandstone, which is but rarely visible on the beach, is seen to be covered by a very beautiful claystone porphyry, containing large crystals of glassy felspar and numerous granular crystals of transparent quartz. It forms a line of cliffs from near Black Water to Corrie Craobhidh, and rises in a mass of rounded hills, called the Leac a Breac, to a considerable height. The blocks on the shore have been quarried for economical purposes; and when split, the influence of the weather may be seen to have altered the colour of the rock, sometimes to the depth of a foot or eighteen inches. Farther north-east, between Slaodridh Water and Clachan Glen, extensive mosses intervene, through which occasionally are protruded blocks of claystone, and in the little streams masses of decomposing syenites and traps.

The district between Leac a Breac and Slaodridh Water is principally porphyritic. At the bridge, however, the bed of the stream is formed of a soft white sandstone, alternating with layers of red micaceous sandstone and coarse shale. A trap dyke about six feet broad traverses these formations (N.E. by E.) above the bridge, and the strata in immediate junction are so much altered, that it is difficult to define the boundaries of the dyke. About a hundred yards above this point, the sandstone gives place to a fine claystone porphyry, which is possibly a continuation of the great mass of Leac a Breac. This rock here forms the bed of the stream for more than a mile, and where the water has worn a channel of sufficient depth, the porphyry assumes a rudely columnar form. In many of the cliffs so exposed, the surface exhibits the appearance of a number of concentric lines surrounding a central nucleus, (see fig. 19,) and this may probably have some analogy with the fact, that the igneous rocks of this class always decompose in a rounded form.

Figure 19.

A little below Glen Rie mill, a white and reddish sandstone crops out from beneath the porphyry, and is again succeeded by porphyry, which has much deranged the order of the stratification, and indurated the sandstone at the junction; so much so, indeed, that the sandstone almost assumes the appearance of a fine grained porphyry. About a hundred yards above the mill, the sandstone re-appears associated with beds of shale, and dipping to the S. W. about 30°. This is one of the most marked instances in Arran where, from the effects of contact with igneous formations, the sandstone has undergone a signal alteration, being highly indurated, and the parallelism of the layers much contorted. The long continued erosion of the stream has cut a deep channel in the sandstone, and on the side of the precipices where the sandstone is indurated, concentric circles may be seen exactly similar to those already noticed in the porphyry. The strata are here composed of beds of yellow and white sandstone, alternating with beds of red shale and soft red sandstone, much interpenetrated by various dykes. These strata are here forced up in terraces (dipping to the S. W. about 29°) in the same manner as the beds of lime at Corrie. It is unnecessary farther to follow the course of the water, which continues to present analogous phenomena to those already described, the hills above being crowned with the various igneous formations which principally compose the entire southern district.

Torlin water, with its accompanying tributaries, exhibits appearances of nearly the same description as those at Slaodridh, the hollows of the streams being composed of beds of shale and highly calciferous sandstone, with occasional thin veins of lime. Numerous trap dykes pierce these strata, and a broad belt of greenstone, resting on sandstone, crosses Smurag water near its source. The surrounding hills are principally porphyritic.

Immediately below Lag Inn an upcast of the various strata has taken place, in which are exposed several trap dykes, which, going from north to south, may be seen in the following order:—1st, A greenstone dyke, thirty feet broad, varying in its direction, but generally pointing to the north. 2d, A greenstone dyke, ten feet broad, (N. 45° E.) and dipping 80° to the west. 3d, A wedge shaped dyke, three or four paces broad, of fine decomposing greenstone, (N. 10° E.) and dipping 80° N.W. And last, to the east of Lag Inn and of the neighbouring distillery, a dyke, twenty-five or thirty feet broad, (N.) dipping to the east about 70°, and crossing the stream on its north bank. One of these dykes below Lag Inn considerably indurates the penetrated strata. On the shore, to the west of South End harbour, is a long group of six dykes named Claitshimoré Point, but only two of them are of any considerable size. Farther west is the singular port called South End Harbour. This port, which is one of the most curious objects on this coast, is entirely natural. Two long dykes—one of which is very thick—the one pointing to the north, the other N. 10° E., form the east and west boundaries of the harbour. A small transverse dyke, stretching from east to west, between the other two, forms a jettee, which shelters the interior of the port from the fury of the waves, and leaves on the west side a wide entry for vessels, which, in this beautiful and regular basin, may lie at anchor in safety from every wind. Lastly, as if to complete the architecture of this basin, a pretty large dyke stretching to the east, bounds the port on the land side, and forms a quay on the sandy beach.

Several other dykes belong to the same group. The natural jettee at the mouth of the port is one of five divergent dykes, pointing between N. 45° E. and E., and which join in a point about N. 70° E. A little farther east is another small dyke, (N. 45° E.,) which, towards its southern extremity, is divided into two branches. The long dyke, also, which forms the eastern side of the port, is accompanied to the west by a parallel dyke thirty feet broad. Between Kilbride Point and the sandy beach below Lag, are two long promontories, formed by vertical dykes pointing N. 10° E. Further west is a shorter promontory, containing one or two dykes, having the same direction as the former.

Between Lag and Benan Head lies an extensive tract of undulating table land. Where the burns cut deep enough into the alluvium, a red sandstone is often laid bare, but the summits of the little rising grounds which here and there dot the surface, are often composed of traps or porphyries. Whether or not an extensive tract of sandstone, intersected by dykes, immediately underlies this alluvium, it is impossible to say, owing to the depth of the soil, and the comparatively level surface of the country; but it often occurs, that in a district of this nature, the alluvial deposits are much intermingled with, or principally composed of, the underlying decomposing rocks. Here the principal embedded pebbles are porphyries and traps, and partly from this reason, partly from the occasional occurence of knolls of igneous formation, it may be, that, like the greater part of the south end of Arran, this tract consists of a large bed of sandstone, which has been penetrated, and in great part overflowed, by trappean and porphyritic formations.

On the coast between Torlin Farm and East Benan is a vast range of cliffs, called the Struye Rocks, which frequently rise to a magnificent height, and assume a rudely columnar form. They are principally composed of a greenstone or basalt, exactly similar to what in German collections is called Anamesite. It contains occasional crystals of hypersthene, which, on the rock being fractured, may be distinguished by a dark metallic lustre. The rocks are frequently intersected by trap veins of a different colour and description from the principal mass. The eastern extremity of the cliffs consists of a beautiful claystone porphyry, which stretches inland in a long low hill, and is apparently a bed overlying the sandstone strata.

It is probable that the carboniferous series here gives place to the new red sandstone, though, from the nature of the ground, it is impossible to define their boundaries. If so, the former must here dip below the latter, which henceforth continues to occupy the coast and the hollows of the valleys the whole way to Brodick Bay. The reasons for this distinction will be better understood when the remainder of the southern district has been traversed.

On the north side of the road, somewhat further east than Benan Head, a beautiful section may be seen, where the stream has cut a hollow in the rocks to the depth of about seventy feet, over which it descends in a fine cascade from a small ravine above. Here a trap dyke penetrates the sandstone formations, and having reached the surface, the melted matter spreads itself over the sandstone strata in all directions to a considerable depth. The horizontality of the underlying sandstone is little disturbed. It is of a dark red colour, and contains small scales of mica. It also contains numerous irregular nodules of fine limestone, varying in dimension from the size of a pea to five or six inches in diameter. Such nodules are of frequent occurence in the new red randstone.

Immediately above the fall, the sandstone is joined by a very compact basalt, slightly porphyritic; and adjoining this rock, and only separated from it by a very small and narrow valley, which constitutes the march of the farms of Levan Currach and East Benan, is a small hill, composed of claystone porphyry, similar to that at Benan Head, and containing a considerable proportion of quartz and felspar. The quartz frequently assumes the form of six-sided crystals, the edges of which are much rounded, and which gives it more of a granular than a crystalline appearance.

To the north-east of the farm of Auchinhew is a most picturesque waterfall, called Eiss a Mor, or the great fall, which precipitates itself over a lofty precipice, into a magnificent amphitheatre, surrounded by perpendicular cliffs, the lower part of which is composed of layers of sandstone, interpenetrated by dykes, which end in overlying masses of greenstone and basalt, partly jointed, and arranging themselves into rude and irregular prisms. Where the sandstone joins the igneous rock, it is much indurated. The parallelism of the layers is slightly discomposed at the penetrating dykes, forming small faults. Similar appearances are visible in Bailemenoch Glen, through which flows the first stream to the east of Eiss a Mor.

On the coast between Benan Head and Druimlaborra, there are twenty-three dykes, stretching N. 10° W. into the sea. One of these is particularly large and elevated, and forms the natural jettee of the Port of Druimlaborra. The two dykes nearest Benan Head are the only small dykes of this group, the eastern one pointing N. 45° E., and the other N. 45° W. Further east is another group of dykes, stretching in a north and south direction. Near Kildonan Castle, there are on the coast ten large dykes (N. or N. 10° W.) which intersect the new red sandstone, and stretch out into the sea. Some of these are elevated, and at the junction, the sandstone is white and indurated. A splendid range of lofty precipices occupies the shore between Kildonan and Learg a Beg, forming a bold and precipitous coast. Like the Benan cliffs, the place is still and solitary. A rough and difficult footpath forms the only track beneath the cliffs; and as the geologist warily winds along, he will hear no sound save the dash of the breaking waves, the shrill cries of the water-fowl, and the incessant cawing of the rooks, which float in airy circles round the verge of the overhanging cliffs. These rocks form a bed overlying the common red sandstone, which may be seen beneath it on the shore about the middle of the range. At another place, the sandstone extends about half way up the cliff. The overlying rocks are almost entirely composed of a dark coloured syenite. At one place there is a mass of fine claystone porphyry, which, however, most probably is a dyke, as it only extends for a few yards in width. Where the syenite joins the sandstone, it assumes a fine compact basaltic texture. From Learg a Beg, which forms the north-east extremity of the coast range, the cliffs stretch inland, gradually rising, and forming one of the series of lofty terraces or steps characteristic of the trap formation, so beautifully marked on the hill to the south of Whiting Bay, and which, as far as Eis a Chranaig, partly form the south side of Glen Ashdale.

The inland district between Auchinhew and Eis a Chranaig, is partly porphyritic and partly basaltic, with scattered patches of other trap rocks, which may either be overlying masses or dykes. Some of the porphyries are so like indurated sandstone, as to be almost undistinguishable from it. Most of the country is, however, covered with moss, so that it is only on the elevated summits, and in a few water courses, that the rocks are visible.

The various appearances exhibited at Eis a Chranaig are exceedingly instructive, as regards the phenomena consequent on the contact of igneous and aqueous formations. The stream which flows through Glen Ashdale, here falls in a splendid cascade, broken only in one place near the summit. The first break is about sixty feet high, and it is possible to descend into an esplanade of rock into which it falls, from whence the second fall makes a sheer descent of great height into the very bottom of the glen. The stratified rocks on the south side of the fall are pierced by a trap dyke about ten feet wide, in connection with which an overlying mass of the same rock has overflowed the sandstone to a considerable depth. The cliffs formed by this greenstone assume an imperfectly columnar form. In the bed of the stream, at the top of the first fall, the igneous formation is larger in the grain than below at its junction with the sandstone, where it is very compact and slightly porphyritic. The sandstone in immediate contact with it is nearly white, exceedingly indurated, and partly assuming a porphyritic aspect, very similar to some of the porphyries in the tract farther south, some of which may possibly be altered sandstones. Below this layer of white sandstone, is another of a brownish-grey colour, also much hardened and slightly porphyritic, in which the layers of stratification are not totally destroyed. Farther down, the sandstone assumes its natural red colour.

On the shore between Kildonan and King's Cross Point, the dykes are as usual well developed. More than forty dykes traverse the coast beds between Kildonan and Learg a Beg, varying in width from one to thirty feet, and in direction from N. 10° E. to N. 20° W. About forty dykes traverse the red sandstone shores of Whiting Bay, some of which intersect each other.

The coast of Lamlash Bay, from King's Cross to Clachland Point, is composed of red sandstone intersected by numerous dykes. On the south side of the bay, three of these are of claystone, and eleven of greenstone. At Gorton Alaster, two of these dykes intersect each other, and a similar instance occurs farther east. They do not preserve any regularity of direction. On the north coast there are eleven dykes all pointing more or less to the north-west.

The alluvial plain of Lamlash is divided on the west into two glens—Glen Alaster and Moneadmhor. The hollows of both these glêns are composed of the ordinary sandstone already so often mentioned, but near the summit of Glen Alaster the sandstone is traversed by a broad vein of fossiliferous limestone, containing the same shells as the strata at Corrie, and which must therefore be part of the carboniferous formations, and probably belongs to a bed of limestone, which, but for the dislocation of the strata might form one of a series of beds superimposed on the calcareous conglomerates of Clachan Glen, which in that case would form the lower beds of the carboniferous series, in the same manner as the çalcareous conglomerates of Corrie and the Fallen Rocks. This carboniferous formation will in this case dip under the new red sandstone somewhere in Glen Alaster, though, from the nature of the interruptions, it may not be possible to ascertain the precise locality of the dip.

Before concluding this chapter it is necessary to assign more fully than has yet been done, the reasons for the classification of the opposite coasts of the southern district with the carboniferous or new red sandstone formations.

By a reference to the map it will be seen that a broad band of old red sandstone crosses from Brodick Bay to the west coast of the island. This can only be identified with the old red sandstone of Glen Sannox by its similarity in structure and composition to that formation, and by the circumstance that both are succeeded by the various alternations of the carboniferous series, the first at Corrie, and the other at the Windmill Hill. As already remarked, it is impossible to define the southern limits of the belt of old red sandstone which crosses the island, as great part of it is probably covered by the alluvial plain through which flow Mauchrie and Black Water; but as already stated, the limestone of Clachan Glen, though belonging to the lower calcareous conglomerate series of beds, probably bears the same relation to this belt as the limestone of Windmill Hill does to the old red sandstone hill which divides Glen Rosa and Glen Shirrag. The relation of these to the fossiliferous lime of Glen Alaster has already been pointed out. The strata laid bare by Slaodridh and Torlin Waters in the south-western district, have already been shown to be composed of calciferous sandstones, shales, &c., similar to what occur in the other districts of the coal measures, and are thus in connection with the limestone strata, considered as belonging to the same formation.

The new red sandstone formations of Brodick Bay are identified with that in the north of Arran by similarity of structure and character, and the marked difference which they present to the principal mass of the underlying carboniferous strata. It is said that a fossil Calamite was found in the sandstone on the shore opposite Springbank, and this is of common occurrence in beds of the lower new red sandstone. The conglomerate at the top of Glen Dubh (Glen Cloy) dips somewhat in the same direction as the carboniferous strata of Windmill Hill which it succeeds, but owing to the intervention of various igneous rocks, its angle of inclination is increased. The conglomerate on the south side of Brodick Bay, though at a lower angle, corresponds in the general direction of its dip with the carboniferous strata on the opposite shore, and has been traced along the Corriegills shore to Clachland Point, and from thence round the eastern coast of the island. This formation must therefore overlie the carboniferous rocks somewhere in the interior of the southern district, but owing to the great overflow of igneous rocks, it is impossible accurately to define their approximate boundaries.

Before concluding, it may be mentioned that the dependency of Holy Isle does not differ in structure from the southern district of Arran, presenting the common red sandstone occasionally visible on the shore, overflowed by a great mass of claystone and claystone porphyry to the height of about a thuusand feet. In the south-eastern district, veins of trap may be seen penetrating the sandstone horizontally.

The Island of Pladda consists entirely of a species of rock intermediate between basalt and greenstone, overlying sandstone, which may be seen for a small space on its eastern coast.

Columnar Rocks of Drummedoon.

Angular Caves near Glen Sannox.

CHAPTER VI.

GENERAL REMARKS AND CONCLUSIONS DRAWN FROM THE OBSERVATIONS MADE IN THE PRECEDING CHAPTERS—THE UNDERLYING GRANITE OF LATER ORIGIN THAN THE SUPERINCUMBENT STRATIFIED FORMATIONS—DIFFERENCE IN POSITION OF THE SOUTHERN IGNEOUS FORMATIONS AND THE NORTHERN GRANITES—EVIDENCES OF IGNEOUS AND AQUEOUS AGENCIES—ERUPTION OF THE SOUTHERN IGNEOUS ROCKS—BOTH CLASSES OF ERUPTED ROCKS NEWER THAN THE STRATIFIED FORMATIONS—THE TRAPS, PORPHYRIES, ETC., NEWER THAN THE COARSE GRAINED GRANITE—DIVISION OF GRANITE INTO THE COARSE AND FINE GRAINED VARIETIES—TRAP DYKES, ETC., IN THE COARSE GRANITE—PROBABLE POSTERIOR ORIGIN OF THE FINE GRANITE TO ALL THE OTHER FORMATIONS—ABRUPT TERMINATION AND DENUDATION OF TRAP DYKES—GRANITE ONCE IN A STATE OF FUSION—PROOFS OF THIS FROM ITS EFFECTS ON THE SLATE AND OTHER PHENOMENA IN CONNECTION WITH THAT FORMATION—ANCIENT ELEVATION OF THE SLATE—QUARTZ VEINS IN THE SLATE—THEIR ORIGIN—UNCONFORMABLE POSITION OF PART OF THE SLATE—RECENT ELEVATIONS.—CONCLUSION.

HAVING thus finished the tour of the Island, and completed a series of observations embracing the principal characteristics of all its various formations, it now only remains to draw certain conclusions which have not been deduced during the progress of the investigations, as they could not all properly be summed up, till these observations were completed.

In the preceding chapter, it has been shown that the whole of the interior of the northern and more mountainous district of Arran is a great mass of granite, forming the lowest rock of the series, and which is again surrounded by various stratified formations, consisting of the hypogene schists and slates, the old red sandstone, the carboniferous series, and the lower beds of the new red sandstone.

This granite often towers above the surrounding rocks in the boldest and most picturesque forms, to the height of nearly three thousand feet. But though highest in topographical position, it has been shown that it is geologically the lowest of the rocks of which the island is composed; and it can also be incontrovertibly proved to be of later origin than any of the superincumbent strata. It is generally found to be the case, that conglomerates or puddingstones are composed of fragments of older rocks in their vicinity; and in Arran, the conglomerates of the old and new red sandstones contain numerous fragments of schists and slates derived from the underlying hypogene strata. But there has not been found in any instance a single fragment of granite in any of the superincumbent strata; from which it may be inferred, that when the stratified rocks, which are many thousand feet in thickness, were formed, the granite had not as yet been exposed to the degrading action of the weather, or the waves; and that if formed at all, it only existed at an unfathomable depth beneath the surface of the stratified formations.

The massive formations in the southern part of the island, form a striking contrast to the plutonic rocks noticed above. It has been shown, that the latter, though newer than the strata which repose on them, occupy the lowest geological position: whereas the igneous formations of the southern district, which are also newer than the stratified rocks, with the exception of the penetrating dykes, which are cut off at the surface, invariably occupy the highest points both geologically and topographically.

Now that the bitterness of controversy has died away, it may well be a theme of wonder, on what ground the geologists of old could lose half a century in useless disputations regarding the comparative merits of their Volcanic and Neptunian creeds. One would certainly now suppose, that a glance at this portion of Arran is sufficient to show that both agencies had long been at work. It has been seen, that except in the sloping banks close on the sea shore, it is generally only where the long continued action of streams has worn away the superincumbent igneous rocks, so as to form deep glens and ravines, that the stratified formations are visible. And when these otherwise hidden strata are so exposed, often in perpendicular faces of great height, they may be seen to be covered by huge masses of traps, porphyries, and syenites, which have burst through, and overflowed the stratified sandstone formations. These overflows in many instances originate in dykes penetrating the strata often at right angles to the plane of stratification; and from the top of these dykes, where the melted mass had liberty of motion in an unresisting medium, whether of air or water we know not, the lava has overflowed the strata on all sides, frequently to a great depth. In such cases, when laid bare in perpendicular faces, it usually assumes a rudely columnar form. From the foregoing remarks, the analogy between these igneous formations and the modern lavas will at once be evident.

In some places where the depth of the superincumbent mass can be examined, these rocks may be about a hundred feet thick, and the sandstone is comparatively undisturbed, retaining its original horizontal position. But it is not on this account to be supposed that this horizontality is always produced right through beneath the superincumbent rocks, occupying the centre of the southern district. In that case the overflowing mass might be often fully eight hundred feet in thickness. But this is not the case, as, in the beds of the streams in the upland glens, the same strata of sandstone are frequently exposed. Such immense overflows do however occur, as in the case of the Holy Isle, where the sandstone on the shore is covered by an unbroken mass of porphyry, nearly a thousand feet in thickness. It will at once be evident, that it could only be a very long continued volcanic action which produced such a large and elevated overflowing mass.

The whole of the south end of Arran must in fact have been formerly in a state of violent volcanic action, but though the igneous rocks are geologically all of the same period, (Section No. 2,) it is not on that account to be supposed that all the eruptions were simultaneous, since what historically may be considered a long period, is, geologically, but a very little time.

From the remarks in the preceding pages of this chapter, it appears that the underlying erupted rocks, or the granites, and the overlying erupted rocks, consisting of traps, porphyries, syenites, &c., are both newer than any of the four stratified divisions which encircle the former, and partly or altogether underlie the latter.

The question thus arises, which of the two classes of igneous rocks is of the latest origin. It was formerly a received opinion that granite was the most ancient of all geological formations, whether igneous or stratified. But recent investigations have shown that this is not the case. Sometimes it is older, sometimes it is newer, than other igneous rocks; and the mere fact, that in Arran the most ancient stratified deposits rest on the granite, (Section No. 1,) is no proof of its superior antiquity to the igneous rocks of the southern district, which rest on the later stratified formations, since it has been shown from the absence of granitic fragments in the conglomerate, that the granite is even newer than the new red sandstone.

A closer inspection of the granitic region, partly solves this difficulty. It has been shown, that there are in Arran two granites, perfectly distinct in their distinguishing characteristics; the one is coarse grained and crystalline, the other of a much finer texture. There are, of course, many minor varieties in these great classes, but the principal divisions are so marked and obvious as to warrant its separation into two distinct classes. These granites are not confusedly intermingled with each other, but occupy, the fine granite the centre of the granitic district, the coarse variety forming the west side of Beinn Mhorroinn and the principal part of the lofty and serrated ridges which form the deep corries and glens on the east coast. Owing to the accumulation of large loose masses of granite which obscure the surface of the rocks, it is impossible to define the actual line of demarcation which bounds the different species, but their approximate boundaries may be traced on the west descent of Beinn Ghnuis and around the back of the long precipitous ridge which extends from thence to the head of Glen Sannox, thence crosses Caistael Abhael and continues to occupy the district nearest the slate to Toirnaneidnoin, and around the western district to the east side of Beinn Mhorroinn. It thus appears that though large grained granite occupies the greater part of the more elevated district, it also occupies much of the lower granitic region, but only where it approaches the slate.* The central district which contains the fine granite, is comparatively low and undulating, arising probably from its greater softness and liability to decay, so that if it ever were topographically as high as the coarse grained peaks, it is possible that the action of the elements in the progress of degradation, has given to the hills their present rounded character.

A considerable number of dykes have been traced in the coarse grained granite, (such as in Section No. 2,) consisting of trap, pitchstone, and porphyries. Thus there are traces of a small greenstone dyke near Brodick mill dam, and of another between that locality and the foot of the steep ascent of Goatfell. Immediately to the north-west of the summit of Goatfell is a claystone dyke occupying a fissure in the rocks. A little above the small bridge near the confluence of Glen Rosa water and Garbh Alt, are two basalt dykes running in an east and west direction, the most northerly being one foot broad and the other fifteen. In the hollow of Glen Rosa, between this point and Ciormhor, there are three porphyritic trap dykes, all stretching nearly north and south.

In the bed of Garbh Alt, near the base of Beinn Ghnuis, is a broad trap dyke closely enclosed between two walls of granite, and forming the bed of the stream for a considerable distance. On the hill on the east side of this glen is a greenstone dyke five or six feet broad, and on the west side of the same glen is another near the foot of Beinn Ghnuis.

Traces of a greenstone dyke may be seen at two places on the northern district of Beinn Ghnuis, and near the summits a greenstone dyke may be seen in contact with the granite. Fragments of pitchstone porphyry are also scattered over both the south and part of the west shoulder of this mountain, but no positive dykes have yet been observed, though there can be little doubt that these fragments owe their origin to dykes now hidden beneath the detached rocks and soil. Further north on the same

* It may be remarked, that when speaking of the hypogene stratified rocks collectively, they are, to save repetition, generally all included under the name of slates.

ridge at the summit of Bealach-a-Nidbhoe is a greenstone dyke, and along the same ridge, pitchstone porphyry frequently occurs in fragments, and in one or two instances in situ. Similar fragments are of frequent occurrence in the descent behind Ciormhor, where there exists a remarkable vein of this mineral, which has already been fully described. There are also two pitchstone porphyry dykes near the summit of Caistael Abhael, and a trap dyke forms, for a short distance, the bed of the stream of Glen Eisnabearradh.

All the dykes above noticed, without a single exception, are contained in the coarse grained granite, and their presence in these localities at once indicates their posterior origin to the rocks which they penetrate, and if these dykes, as is probable, are contemporaneous with the *traps*, &c. of the south end, then *this granite must also be older than they are.* But in the central fine grained granite to the west of this elevated region, none of these appearances are manifested; and as far as this district has been examined, no dykes of the kinds above indicated have yet been found to penetrate it. It ought to be stated that the Beinn Mhorroinn district, as regards these appearances, has not yet been investigated; and in the event of phenomena being there discovered contrary to those now described, it may partly or altogether invalidate the conclusions now about to be drawn. Now, as in cutting through the coarse granite, these dykes frequently approach its borders where it joins the fine variety, and are there invariably cut sharply off by the fine granite, when they approach it in the coarse kind, it may be asked, has not the fine central granite been protruded posterior to the intrusion of the dykes? in which case had the dykes been prolonged into the space now occupied by the fine granite, its protrusion must have cut them sharply off as they now appear; and, therefore, if as has been already stated, these dykes be of the same age as the southern igneous formations, *then the fine central granite must also be of later origin than they are, and consequently also newer than the external coarse grained granite.* That the fine granite is newer than the last named rock, we have this additional proof, that it is often found penetrating it in veins, varying from an extreme minuteness to many feet in thickness.

It will be remembered that there is a mass of fine grained granite to the west of Glen Cloy, (see Map, and Section No. 2,) associated with syenites, porphyries, &c., and from the similarity of its appearance, and the identity of its mineral character, with many of the fine granites of the interior of the northern granitic district, it seems to be at the least extremely probable that they are of the same age, and in fact belong to the same granite, which has burst through different formations at the same time; in which case, *both of the fine granites must be newer than any other formation in Arran, whether stratified or erupted.*

Before leaving the subject of the trap dykes, it may be remarked with regard to their appearance both in the stratified formations, and the granite, that it cannot be supposed that they terminated so abruptly (see Section No. 2,) without overflowing the penetrated rocks, similarly to what has been occasionally pointed out, as occurring in the stratified formations in the trappean districts of the south end. We may well then be astonished at the vastness of that denudation, which has entirely worn away all traces of the overflowed masses of traps and porphyries, except when the abrupt termination of the dyke is seen on a level with the penetrated formations.

When treating of the junctions of the granite and slate, it will have appeared that the granite in Arran is considered to have been at one time in a state of fusion, an opinion which, considering the analogy between the plutonic rocks and the various formations of the trap family, is now pretty generally received. But it may be asked on what authority this conclusion is drawn respecting the Arran granite.

It has already been shown that the granite exhibits every symptom of having been intensely heated when it first touched the slate; first, from the partial fusion of the slate in some localities when in contact with the granite, and, secondly, by the violent contortions in the stratification of the slate wherever it approaches the granite; and this heat must have been of long continuance, since these contortions are often very great, at the distance even of a mile and a half from the granite, as exemplified at North Thundergay, and in various other localities. That the granite was also at least partially melted, is sufficiently clear from the numerous granitic veins injected into the schists and slates. But on examining the junction where such interpenetrations occur, the difficulty is at once suggested,—if the granite were in a state of fusion when it broke through the strata, why, besides penetrating the slate in veins, does it not also overflow it? Had it been in such a state when raised to its present elevation, it must inevitably have overflowed, like modern lavas, or the traps and porphyries of the south end of Arran.

The following seems to be the only method of accounting for this phenomenon. Probably before the deposition of the old red sandstone, the melted granite was at some distant period formed under the slate. The effects of the intense heat acting on the slate, produced those violent contortions already so frequently noticed. The slate also seems, from the same cause, to have been much cracked or fissured on its lower surface. Into these fissures the granite infused itself, there cooled and solidified, (fig. 22,) and was subsequently by other agencies upheaved

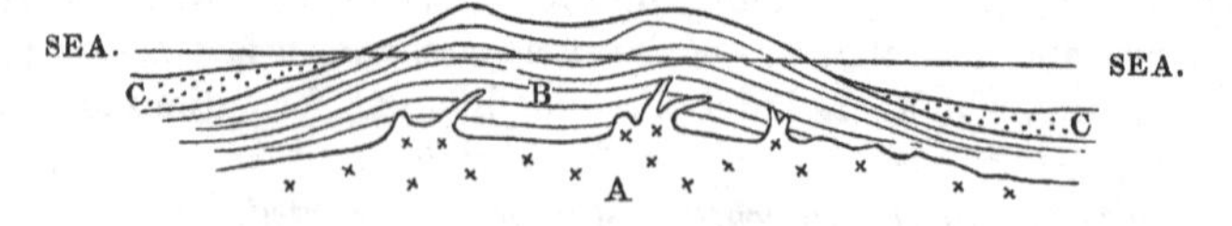

Fig. 22. A Granite. B Clayslate. C Old Red Sandstone Conglomerate.

and protruded in its present crystalline form, many of the veins still retaining their position in the slate, (fig. 23.)

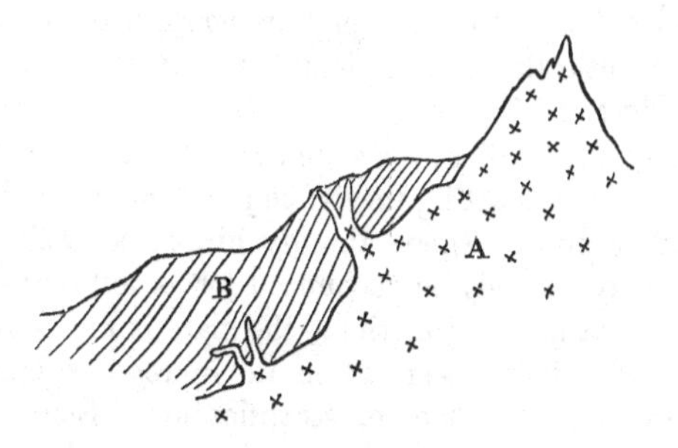

Fig. 23. A Granite. B Slate.

The protrusion of the granite in a solid form, is also borne out by the manner in which the stratified rocks incline against it, which is generally at a very high angle. This is not generally the case with the igneous formations of the south end of Arran, where the traps, porphyries, &c., break through the stratified formations, often without producing any material derangement in the level of stratification.

A partial elevation of the slate seems also to have attended the first injection of the granite, long previous to its final protrusion. Two striking circumstances increase the probability of this first partial upheaval. First, that innumerable waterworn fragments of the identical slate and schist are found embedded in the old red sandstone; and on its upper surface, which is many hundred feet above its junction with the slate, the fragments are as numerous as they are at its lower surface, which could not have taken place unless the hypogene strata had previously been elevated to a considerable height above the sea, after which, either by the waves on the coast, or by streams, they were partly worn into sand and shingle, now forming one of the ingredients of the conglomerate which reclines against it, (fig. 22.)

The second evidence of this ancient elevation is this,—that in many districts of Arran, a shallow hollow intervenes between the highest point of the slate and the steep ascent of the granitic ridges. This appearance was particularly adverted to, in the description of the district lying between Brodick mill dam and Maoldon. From this appearance, it may be inferred, that when the solid mass of underlying granite, from whatever cause, was forced through these strata, the separated edges would dip towards the granite in the form shown in figure 24.

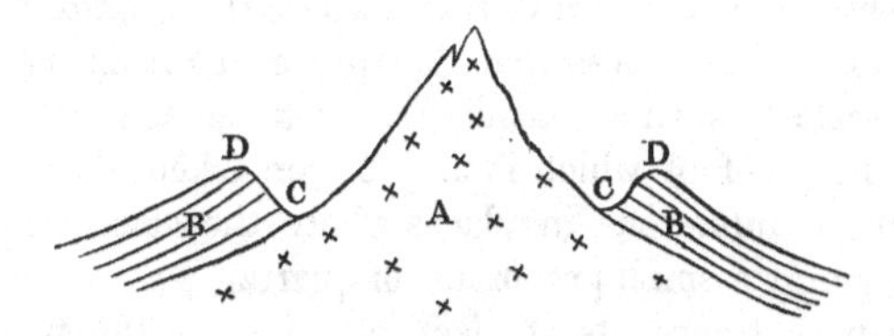

Fig. 24. A Granite. B B Superincumbent Slate. CC Intervening Shallow Valleys. D D Slaty Knolls.

But had the old red sandstone completely overlaid the slate previous to this elevation, it naturally, unless denuded, might have occupied the place of the low ridges, D. D.

The contortions of the strata of the slate have already been sufficiently treated of in the notices respecting its visible junctions with the granite: but there is another phenomenon, already hinted at, well worthy of notice, and which may possibly be accounted for as follows:—

It was observed, that at the junctions, the slate is much veined with quartz. These, it has been seen, often lie in regular laminæ, very numerous and parallel to the plane of stratification. Sometimes these alternations are almost as minute as the leaves of a closed volume, while they are also often found on a very large scale. In other instances they penetrate the slate laterally, being occasionally even two feet in thickness. But this is not confined to the neighbourhood of the granite alone. On the contrary it has been found that both interlaminations and large lateral veins, occur at great distances from the granite, as on the coast near Thundergay, and on the east side of Loch Ranza. These laminar veins are not universally diffused throughout the strata, and in many large districts, distant from the granite where the strata are comparatively undisturbed, veins of quartz are much more rare, but wherever the strata are contorted, however distant from the granite, there numerous veins are developed, or otherwise it assumes a granulated texture, exhibiting numerous grains of quartz, which sometimes partly assume a rounded crystalline form, such as takes place on part of the hills above Lagan and Lagantuin. It is not impossible that the larger veins may sometimes be owing to the infiltration of siliceous particles into fissures, but we must look to other causes for the origin of the smaller veins, and the regular minute quartz laminæ, alternating with similar laminæ of slate. It is evident that the interlamination of quartz, has nothing to do with the original deposition of the strata, for besides that the nature of quartz in continuous crystallized veins precludes this supposition, the fact of many of them running at various angles to the plane of stratification, is of itself sufficient evidence, that such was not always the case. Neither, even supposing the water to be highly charged with silica in which the strata were deposited, can it be supposed possible, that the siliceous matter would regularly deposit itself in layers alternating with the other mechanical precipitates: it would rather have become blended with them, and formed a cement to bind the particles together.

Now though these veins are always found where the granite and slate join, yet it is impossible to suppose that they proceeded from the granite, and pierced the strata to such a considerable distance as they appear in many instances; in proof of which it may be remarked, that the veins are sometimes more numerous in places where the slate reclines on granite, containing only a small proportion of quartz.

It is well known that currents of electricity possess the property of separating and arranging intermingled substances. The following instance will best illustrate this property. A quantity of clay and sand were mixed together in a glass tube, along with as much water as merely damped the substances, leaving the ingredients of such a thick consistence that neither of them could be precipitated to the bottom. Through this tube a current of electricity was passed, and in process of time, the clay and sand were found to be separated into two distinct strata occupying different portions of the tube. This agency might therefore have had a powerful effect in the productions of the quartzoze veins and laminæ. But another great assisting cause seems to have been called into operation, namely, the agency of heat.

It is evident, in relation to the effects produced by the heated granite on the slate, that there is a connection between the contortions of the strata, and the development of quartz veins and laminæ, since these veins are comparatively rare in undisturbed districts; and their existence having been shown to be alien to the original deposition of the strata, the most obvious method of solving the difficulty, in conjunction with that already stated, seems to be as follows:—

It is generally supposed that the consolidation of granitic masses took place at great depths in or below the bed of the sea, and consequently under an immense superincumbent pressure. It has already been shown that the granite of Arran is newer than the superincumbent strata, and it was also proved that it was in a state of fusion while underlying the slate. In these circumstances the cooling process would necessarily be exceedingly slow, and even after consolidation, the granite would long retain a high temperature, the more so that it received fresh accessions of heat, in proof of which we have only to refer to the numerous veins of newer granite which penetrate the older rock. In the eruptions of the more modern traps, &c., we do not find the effects of heat visible in the strata at any great distance from the point of eruption, and this may be owing to the comparative rapidity with which the erupted mass cooled. But in the instance of the erupted plutonic rocks of Arran, when the interior heat may have endured for an indefinite period, it need not occasion surprise to find it acting powerfully on strata, at a great distance from the heated mass. If the slate then were so intensely heated, as by expansion to allow its component particles to revolve on their axes* with greater freedom than they could possibly do in a more condensed form, then, by virtue of that law by which kindred atoms are attracted by and cohere to each other, the siliceous particles in the slate would in the process of time congregate together, forming veins wherever circumstances most easily allowed, which would naturally be between the layers of stratification, where they most commonly occur; and this separating process would, as before stated, be materially aided in its operations from the influence of electric currents consequent on the development of so much caloric.

There still remains another point to be discussed with regard to the

* It is well known that fluidity is not absolutely essential to allow particles to revolve on their axes. It is also evident that increased expansion will materially aid the process.

slate; viz, the relation of the dip of the strata to the granitic centre. Though subject to many variations and minor disturbances, the principal portion of the slate may be said to be conformable to the mineralogical centre. From Glen Sannox round the south side of the granite, and on the west coast by Beinn Mhorroinn, the slate in general reclines against the granitic mountains. On the north coast between Catacol and Loch Ranza, the strata dip towards the granite, but the plane of stratification conforms to the central nucleus, and therefore the strata cannot be said to be unconformable to it. There is, however, one most decided exception to this rule, in that district lying between North Sannox and Newton Point. On the hills to the south of the junction of Glen Dubh and North Sannox, the edge of the slate is seen to abut against the granite, and though from the nature of the ground it is impossible to determine how it joins the granite between Glen Dubh and Toirnaneidnoin yet the long ridge of schistose hills from Lagantuin to the Cock, uniformly dip to the southwards, at an average angle of about 74°. This is then quite unconformable to the mineralogical centre. But it has been seen that part of the old red sandstone, the carboniferous series ranging along the shore from the Fallen Rocks to the north of the Salt Pans, and the new red sandstone above these, are all obedient to the dip of the anticlinal line, and so far conformable to the mineralogical centre, resting on the elevated schists intervening between them and the granite at various angles of inclination dipping to the north and east. The schist and the newer formations, therefore, exhibit here an instance of unconformable stratification, (Section No. 1.) A well known instance of this has been described as occurring at Newton Point, where a small patch of new red sandstone rests on the upturned edges of the strata, (fig. 12, page 36.)

In this instance it is perfectly clear that the edges of the slate must have been upturned previous to the deposition of the overlying sandstone; and probably the hypogene strata of this entire district were thus disturbed previous to the subsequent depositions. There is however another hypothesis, which though it cannot be applied in all the circumstances of this locality, may yet possibly bear some relation to this very apparent unconformability.

A reference to the map will show that a prolongation of the plane of stratification brings the west edge of the slate in contact with the granite, but not at right angles to it. In fact, the line of junction and the average plane of stratification from Toirnaneidnoin to Suithi Fheargus make an angle of about 40°, so that the slate though unconformable with the granite to the west, is always partly conformable with the great granitic ridge to the south. Draw a line north and south from the junction at Suithi Fheargus to the point between the Fallen Rocks and the Salt Pans, and the plane of stratification which there runs east and west, is perfectly conformable with the point from whence the line is drawn. Seeing then, that the plane of stratification is conformable to the principal granitic mass of the district, though the strata lean from the granite, it may perhaps be *possible*, that the protrusion of the granite tilted the upper edge of the slate beyond the vertical line, without producing the same effect on the topographically lower, though geologically superincumbent strata. As the slate was elevated above the sea previous to the deposition of the sandstones, this might indubitably have taken place in the manner shown in figure 25, though it does not clear up the difficulty respecting the unconformable slate at Newton Point and at the Cock, which would appear to have entirely resulted from the previous derangement of the hypogene strata. This, therefore, is a part of the geological history of Arran which requires farther investigation.

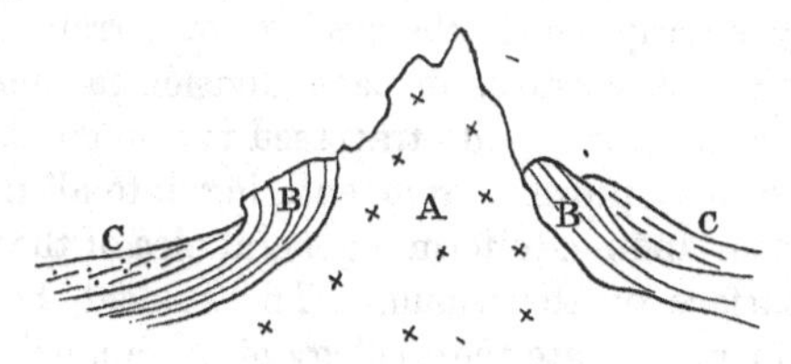

Fig. 25. A Granite. B B Slate. C C Superincumbent secondary formations, apparently partly unconformable to the subjacent slate.

It still remains to make a few remarks on the more recent elevations of the island. It has already been shown that an ancient elevation of the slate occurred previous to the deposition in any of the secondary strata. It might perhaps be possible to discover signs of many other partial elevations, but they have not hitherto been discovered. It has been shown that an ancient sea cliff surrounds almost the entire island, the bottom of which averages perhaps about forty feet above the level of the sea; and that between this cliff and the sea is an ancient beach containing many shells, which are found even at the entrance of the water-worn caves so common in this cliff. It has also been remarked that these caves, page 61, dip to the south agreeably to the inclination of the anticlinal line, their pillars being at right angles to the plane of stratification, not to the horizontal level; proving that their formation by the influence of the waves, took place previous to their elevation to their present height. But independently of this conclusion, this observation determines a most important point, which clearly indicates the geological period of the last upheaval of the granite:—

These caves were excavated by the sea at the time when the recent shells of the raised beach were deposited within them. They are therefore of recent origin. It has been shown that the strata dip in both directions from the anticlinal axis, conformably with the granitic centre. The anticlinal axis is therefore caused by the protrusion of the granite, and as the water-worn caves dip agreeably to the anticlinal line, their pillars being at right angles to the plane of stratification, it follows *that an elevation of the granite occurred after their formation, or in other words, when shells such as now exist were common on the shore.*

There is still great room for further investigation in Arran.* It is desirable that the unconformability of the slate and schist on part of the eastern coast be completely explained; that all the relations of the two granites be more thoroughly investigated, including the extent of surface occupied by the granite of Ploverfield, and its connection with the surrounding igneous and stratified rocks. It may also be possible to determine the relative ages of the various overlying igneous formations; and it is of importance that the secondary stratified formations, more particularly in the southern district, be examined, in the hope of discovering fossils.

In the preceding description of the geology of Arran, it has been studied so to divide the chapters, as in each division to comprise a district of such extent, as may be rapidly traversed in a single day, by those whom circumstances may prevent from entering into all the minutiæ, but who wish, notwithstanding, to form a general idea of the phenomena exhibited on the surface of the island. The student, however, who desires thoroughly to investigate the geology of Arran, will require to conduct his operations in a more leisurely manner; frequently returning to re-examine portions of the country which he has previously gone over, minutely noting every little particular, and comparing and vivifying his observations, till he thoroughly understand the relations of the various formations to each other in every district; and day by day, foul weather or fair, having at last traversed every hill, and followed the windings of every stream, he will finally be enabled to form a clear and correct idea of the entire conformation of the country.

It happens that the inns, and other places where accommodation may be obtained, are so situated around the island, that, by remaining a few days at each, the geologist may arrange his excursions in such a manner, that he will rarely require to traverse the same ground twice, unless he specially desire to do so. The traveller may thus take up his abode at Brodick, Corrie, and Loch Ranza. Lodgings may be got at Dugarry. There are also inns at Shedog, Black Water Foot, at Lag, and at Lamlash. The greatest distance destitute of such accommodation lies between the two latter places. But should the traveller feel weary, and become faint by the way, he will find wherewith to refresh and re-invigorate himself beneath the humble roof of the little auberge at Kildonan.

These, however, are contingencies with which the geologist has nothing to do, and about which he must not trouble himself. Let him breakfast

* The works hitherto published containing information on the Geology of Arran, are, the valuable paper already noticed, written by Professor Sedgwick and Mr. Murchison—Dr. M'Culloch's Western Isles—Professor Jameson's Mineralogical Tour—M. Necker on the Trap Dykes—and Mr. Headrick's work, which, however, is now completely antiquated.

early, and start in the morning with as much provision in his knapsack as will sustain his strength till dark: and when nature begins to fail, if the day be fine, let him stretch himself out to rest on the grass or the heather, by the margin of some clear mountain stream, and there partake of his humble fare: if the day be wet and boisterous, he may eat as he walks, or crouch for a few minutes under the lea of a projecting rock. For his equipments, let him be roughly and strongly clad. He will also require a stout and capacious knapsack to hold his specimens, a note-book, a sketch-book, a good hammer, round and heavy at one end, and having the opposite extremity constructed after the manner of those used in breaking stones for the Macadamising of roads. He will likewise need a chisel and a clinometer, and a broad leathern belt strapped around his waist, in which to thrust his hammer, when, in clambering on the mountain sides, he is sometimes forced to betake himself to the use of both hands and feet. To these, some geologists consider a flask an essential concomitant, asserting that it is indispensable as a preservative against the evil effects of bad weather, and the moist atmosphere of the west Highlands. It is difficult to advise on a matter so delicate; this, however, is certain, that if a guide be employed, the flask will, in nine cases out of ten, be considered by at least one of the party as a most desirable accompaniment.

Contorted Schist.

SYNOPSIS

OF

THE ROCKS OF ARRAN

IN

THE ORDER OF THEIR SUPERPOSITION.

GRANITE.

COARSE GRAINED GRANITES.—Goatfell, and the principal part of the Granite Mountains on the east and west coasts.—Also, in boulders on the shore between Brodick and Glen Sannox, &c.—These generally indicate the description of the granite of the hills above, though it is unsafe to consider the rule invariable.—Coarse grained granite is in immediate contact with slate, at the junction in Glen Dubh.

FINE GRAINED GRANITES IN MASS.—The interior of the granitic district, between the Beinn Ghnuis ridge, and Beinn Mhorroinn.—Ploverfield, to the west of Glen Dubh (Glen Cloy).

FINE GRAINED GRANITE IN VEINS PENETRATING THE COARSE GRANITE.—Great vein crossing the ridge between Ceum Na Cailleach and Siuthi Fheargus.—Small veins in Goatfell and the other granitic hills on the east coast, &c., particularly above Corrie.—Also in boulders on the shore.—Fine granite in contact with, and penetrating, slate or schist, at all the visible junctions; viz., at the Mill Dam, Goatfell—at the White Water, where it is a compact felspar—in the burn below Ciodh Na Oigh—in North Sannox—Glen Dubh—Toirnaneidnoin—Mathaic Uan—Maol Na Leaca Sheanhain, &c.

MINERALS IN THE GRANITE.

QUARTZ CRYSTALS.—Principally in cavities in the coarse grained granites; as on Goatfell, Caistael Abhael. Though these cannot be said to be rare, they are not of such frequent occurrence, as to be often found of any considerable size or perfection.

FELSPARS.—Found under the same circumstances as the crystals of quartz. Said to be not uncommon in Goatfell, Ciormhor, and Glen Sannox.

MICA.—Large transparent plates are not found in Arran. It principally occurs in small brown or black scales. Sometimes the crystals attain to the size of half an inch.

CLAYSLATE AND SCHISTS.

These rest immediately on the central granite.

CHLORITE SCHIST.—Newton Point, and great part of Glen Chalmadael—Slate Quarries.

MICACEOUS SCHIST.—Glen Chalmadael—part of the district between Loch Ranza, and Catacol.

CLAYSLATE OR ARGILLACEOUS SCHIST.—In the upper part of Cnocan Burn, and generally in the higher district of the low hills between Brodick and North Sannox.—Also part of Glen Chalmadael—as at the Slate Quarries near Loch Ranza—at the west point of Loch Ranza—at North Thundergay—and in various small patches on the west coast.

SCHISTS AND SLATES CONTORTED OR OTHERWISE ALTERED BY HEAT.—At, and in the neighbourhood of, all the junctions mentioned above—as at the Mill Dam—Glen Dubh.—Also at Newton Point—North Thundergay.

OLD RED SANDSTONE.

This formation in a sandstone and conglomerate form occupies the east coast, from the Fallen Rocks to the March of Achab Farm near Corrie; and crosses the island from above Brodick to the west coast, from the entrance to Glen Iorsa, southwards. The pebbles embedded in the conglomerate are quartz, slates and schists, jasper, &c. It may be seen alternating with, and merging into the slaty veins in Cnocan Burn and North Sannox water, &c. Reticulations in the sandstone, in the wood at the back of Glen Rosa Cottage.

> SULPHATE OF BARYTES.—In a vein in Glen Sannox.—Also in the bed of the stream which flows into Glen Sannox from Corrie Na Chiodh.

CARBONIFEROUS SERIES.

Rests conformably on the old red sandstone. It joins it at the Fallen Rocks, and forms the coast to where it joins the new red sandstone near Cock, dipping to the north-east in accordance with the anticlinal line, and consequently rests unconformably on the slate and schist which form the higher inland hills. It also occupies the coast from Corrie to Brodick, and forms the south side of Glen Shirrag—King's Hill—Clachan Glen, and part of the hills to the south of the Shiskin Road, and the stratified rocks further south on the west and part of the south coast. Fossil plants (stigmarias, calamites, &c.) in the sandstone at Corrie.

FOSSILIFEROUS LIMESTONE.—At Corrie—Locherim—Maoldon—Castlewoods—Glen Shirrag—Glen Alaster.—Black mountain limestone between the Fallen Rocks and the Salt Pans—Red limestone to the north of the Salt Pans.

UNFOSSILIFEROUS LIMESTONE.—*Calcareous Conglomerate*, resting on the old red sandstone at the March of Achab Farm—immediately to the north of the Fallen Rocks—Clachan Glen—*Cornstone and various coloured limestones, including grey marble,* between the March of Achab Farm and Corrie.

TRANSPARENT CRYSTALS OF CARBONATE OF LIME.—Associated with limestone as at Corrie—Clachan Glen, &c.

SULPHATE OF BARYTES.—Sparingly associated with the limestone of Clachan Glen.

SHALES.—*Variegated Shales at Corrie*—Locherim—Maoldon—Castlewoods—Glen Shirrag—Glen Alaster—of frequent occurrence throughout the carboniferous series to the north of the old red sandstone. Various other coloured shales associated along with the same beds, such as *black shale* along with the black mountain limestone.—*Fossiliferous Bituminous shale* below and above the coal. Various *coarse shales* in the streams on the west and south coast beyond Black Water foot.

COAL.—At the Salt Pans.

HEMATITE or Red Oxide of Iron.—In the shale at Corrie Lime Quarry—on the shore a little to the north of the Fallen Rocks—between the northern bed of red limestone and the new red sandstone.

NEW RED SANDSTONE.

SANDSTONE AND CONGLOMERATE, resting conformably on the carboniferous series, and uncomformably on the slate and schist, from where it joins the former northwards to Alt Mhor.—Unconformably on the schist, at Alt Beithe, Newton Point.—From Glen Cloy southwards along the east coast—part of the south coast. Its structure is similar to the old red sandstone.

QUARTZ ROCK, on the hill immediately to the south-west of Glen Dubh, (Glen Cloy).

OVERLYING IGNEOUS ROCKS.

CLAYSTONE PORPHYRY.—On the top of the Windmill Hill—Dundubh—part of Drummedoon Point—Leac a Breac—Slaodridh Water—Benan Head, &c.

SYENITE.—Light coloured Syenite, on the hills to the west of Glen Dubh, (Glen Cloy). Dark Syenite.—Dunfion and the hills from Dundubh to Clachland Point.—The greater part of the Dippin Rocks, &c.

PITCHSTONE AND PITCHSTONE PORPHYRY.—In Glen Cloy, on the west side of the road immediately to the south of the wood between Glen Shirrag and Glen Cloy—Old Lamlash Road—Corriegills Shore—Beinn Ghnuis ridge—Ciormhor—King's Hill—Tormor, &c.

BASALT.—Part of the mass of trap on the shore between Corrie and the March of Achab Farm—several of the trap dykes, such as the dykes in Garbh Alt near its confluence with the Rosa—the fourth dyke south from the Fallen Rocks on the shore. The first dyke to the south of the new red sandstone quarry near Cock—ditto, to the north—dykes at the unconformable new red sandstone, at Newton Point. The term basalt has been so often used in a vague manner, that it is difficult always to draw a distinction between this rock and greenstone. Here it is used to designate an extremely compact rock of that description. The lower part of overlying greenstone masses sometimes assumes this form, where it comes in contact with the stratified formations.

GREENSTONE.—Most of the numerous trap dykes are of this formation—as those on the coast between Brodick and Corrie. It may be seen in mass at Eis a Chranaig.

PORPHYRITIC TRAP, including greenstone porphyry, &c.—In Cnocan Burn, containing crystals of glassy felspar—dykes immediately north of Fallen Rocks—elevated dyke near Cock, containing crystals of opaque felspar—Drummedoon Point, containing crystals of glassy felspar, and —Amygdaloidal greenstone porphyry,—dykes immediately to the south of Drummedoon Point, containing glassy felspar and carbonate of lime.

AMYGDALOID.—On the shore between Corrie and the March of Achab Farm—in the bed of the first stream to the north of Maoldon—immediately to the north of the Fallen Rocks, &c.

In the igneous mass on the shore between Corrie and the old red sandstone, are small veins of STEATITE, associated with veins of QUARTZ and CARBONATE OF LIME.

ALLUVIUM. (See the map.)

BELL AND BAIN, PRINTERS, GLASGOW.

Printed in the United States
By Bookmasters